On and Off-Chip Crosstalk Avoidance in VLSI Design

Chunjie Duan · Brock J. LaMeres · Sunil P. Khatri

On and Off-Chip Crosstalk Avoidance in VLSI Design

Springer

Authors
Dr. Chunjie Duan
Mitsubishi Electric Research
Laboratories (MERL)
201 Broadway
Cambridge, MA 02139
USA
duan@merl.com

Dr. Brock J. LaMeres
Montana State University
Dept. Electrical & Computer Engineering
533 Cobleigh Hall
Bozeman, MT 59717
USA
lameres@ece.montana.edu

Dr. Sunil P. Khatri
Texas A & M University
Dept. Electrical & Computer Engineering
214 Zachry Engineering Center
College Station, TX 77843-3128
USA
sunilkhatri@tamu.edu

ISBN 978-1-4419-0946-6 e-ISBN 978-1-4419-0947-3
DOI 10.1007/978-1-4419-0947-3
Springer New York Dordrecht Heidelberg London

Library of Congress Control Number: 2009944062

Printed on acid-free paper

Springer is part of Springer Science+Business Media (www.springer.com)

Preface

One of the greatest challenges in Deep Sub-Micron (DSM) design is inter-wire crosstalk, which becomes significant with shrinking feature sizes of VLSI fabrication processes and greatly limits the speed and increases the power consumption of an IC. This monograph presents approaches to avoid crosstalk in both on-chip as well as off-chip busses.

The research work presented in the first part of this monograph focuses on crosstalk avoidance with bus encoding, one of the techniques that effectively mitigates the impact of on-chip capacitive crosstalk and improves the speed and power consumption of the bus interconnect. This technique encodes data before transmission over the bus to avoid certain undesirable crosstalk conditions and thereby improves the bus speed and/or energy consumption. We first derive the relationship between the inter-wire capacitive crosstalk and signal speed as well as power, and show the data pattern dependence of crosstalk. A system to classify busses based on data patterns is introduced and serves as the foundation for all the on-chip crosstalk avoidance encoding techniques. The first set of crosstalk avoidance codes (CACs) discussed are memoryless codes. These codes are generated using a fixed code-book and solely dependent on the current input data, and therefore require less complicated CODECs. We study a suite of memoryless CACs: from 3C-free to 1C-free codes, including code design details and performance analysis. We show that these codes are more efficient than conventional crosstalk avoidance techniques. We discuss several CODEC design techniques that enable the construction of modular, fast and low overhead CODECs. The second set of codes presented are memorybased CACs. Compared to memoryless codes, these codes are more area efficient. The corresponding CODEC designs are more complicated, however, since the encoding/decoding processes require the current input and the previous state. The memory-based codes discussed include a 4C-free code, which requires as little as 33% overhead with simple and fast CODEC designs. We also present two general memory-based codeword generation techniques, namely the "code-pruning"-based algorithm and the ROBDD-based algorithm. We then further extend the crosstalk avoidance to multi-valued bus interconnects. The crosstalk classification system is first generalized to multi-valued busses and two ternary crosstalk avoidance schemes are discussed. Details about the ternary driver and receiver circuit designs are also presented in the monograph.

Advances in VLSI design and fabrication technologies have led to a dramatic increase in the on-chip performance of integrated circuits. The transistor delay in an integrated circuit is no longer the single bottleneck to system performance as it has historically been in past decades. System performance is now also limited by the electrical parasitics of the packaging interconnect. Noise sources such as supply bounce, signal coupling, and reflections all result in reduced performance. These factors arise due to the parasitic inductance and capacitance of the packaging interconnect. While advanced packaging can aid in reducing the parasitics, the cost and time associated with the design of a new package is often not suited for the majority of VLSI designs. The second part of this monograph presents techniques to model and alleviate off-chip inductive crosstalk. This work presents techniques to model and improve performance the performance of VLSI designs without moving toward advanced packaging. A single, unified mathematical framework is presented that predicts the performance of a given package depending on the package parasitics and bus configuration used. The performance model is shown to be accurate to within 10% of analog simulator results which are much more computationally expensive. Using information about the package, a methodology is presented to select the most cost-effective bus width for a given package. In addition, techniques are presented to encode off-chip data so as to avoid the switching patterns that lead to increased noise. The reduced noise level that results from encoding the off-chip data translates into increased bus performance even after accounting for the encoder overhead. Performance improvements of up to 225% are reported using the encoding techniques. Finally, a compensation technique is presented that matches the impedance of the package interconnect to the impedance of the PCB, resulting in reduced reflected noise. The compensation technique is shown to reduce reflected noise as much as 400% for broadband frequencies up to 3 GHz. The techniques presented in this work are described in general terms so as not to limit the approach to any particular technology. At the same time, the techniques are validated using existing technologies to prove their effectiveness.

The authors would like to thank Ericsson Wireless Communications in Boulder, Colorado, Mitsubishi Electric Research Laboratories in Cambridge, Massachusetts and Agilent Technologies in Colorado Springs for funding some of the research works presented in this monograph. Agilent Technologies also provided instrumentation, EDA tools, and hardware used in the development and analysis of the off-chip crosstalk avoidance techniques in this monograph. The authors would also like to thank Xilinx Corporation for providing the FPGA devices and design methodologies necessary for prototyping the techniques in this monograph and evaluating their feasibility.

August 2009

Chunjie Duan
Brock J. LaMeres
Sunil P. Khatri

Acknowledgements

Dr. Duan is grateful to his parents, Derong and Xingling, and his beautiful wife, Rui for their unconditional support over the years. Without their support, research would have been a lot more painful experience than it already is.

Dr. LaMeres would like to thank his family for all of the support they have given over the years. Endless thanks for offered to his wonderful wife, JoAnn, and to his two precious daughters, Alexis and Kylie, who have given up too many nights and weekends of family time for the pursuit of research in this area. Their sacrifice will always be remembered.

Dr. Khatri would like to thank his family for their support and encouragement over the years, without which this book and many other research endeavors would simply not have been possible.

Contents

Abbreviations

ASIC	Application Specific Integrated Circuit
BGA	Ball Grid Array
CAC	Crosstalk Avoidance Code
CMP	Chip-level Multiprocessing
CMOS	Complementary Metal-Oxide-Semiconductor
CODEC	Encoder and Decoder
DSM	Deep Submicron
FC	Flip Chip
FPF	Forbidden Pattern Free
FPGA	Field Programmable Gate Array
FNS	Fibonacci-based Numeral System
FTF	Forbidden Transition Free
LSB	Least Significant Bit
MSB	Most Significant Bit
NoC	Network on Chip
PAM	Pulse Amplitude Modulation
PDP	Power-Delay Product
PCB	Printed Circuit Board
PWB	PrintedWiring Board
PLA	Programmable Logic Array
QFP	Quad Flat Pack
ROBDD	Reduced Ordered Binary Decision Diagram
SoC	System on Chip
SoI	Silicon on Insulator
VLSI	Very Large Scale Integrated Circuit
WB	Wire Bond

List of Figures

List of Tables

Part I
On-Chip Crosstalk and Avoidance

Chapter 1
Introduction of On-Chip Crosstalk Avoidance

1.1 Challenges in Deep Submicron Processes

The advancement of very large scale integration (VLSI) technologies has been following Moore's law for the past several decades: the number of transistors on an integrated circuit is doubling every two years [4] and the channel length is scaling at the rate of 0.7/3 years [41, 68]. It was not long ago when VLSI design marched into the realm of Deep Submicron (DSM) processes, where the minimum feature size is well below 1 μm. These advanced processes enable designers to implement faster, bigger and more complex designs. With the increase in complexity, System on Chip (SoC), Network on Chip (NoC) and Chip-level Multiprocessing (CMP) based products are now readily available commercially.

In the meanwhile, however, DSM technologies also present new challenges to designers on many different fronts such as (i) scale and complexity of design, verification and test; (ii) circuit modeling and (iii) processing and manufacturability. Innovative approaches are needed at both the system level and the chip level to address these challenges and mitigate the negative effects they bring.

Some major challenges in DSM technologies include *design productivity, manufacturability, power consumption, dissipation* and *interconnect delay* [68, 93]. High design cost and long turn-around time are often caused by the growth in design complexity. A high design complexity results from a growth in transistor count and speed, demand for increasing functionality, low cost requirements, short time-to-market and the increasing integration of embedded analog circuits and memories. Poor manufacturability is often a direct result of reduction in feature size. As the feature size gets smaller, the design becomes very sensitive to process variation, which greatly affects yield, reliability and testability. To address these issues, new design flows and methodologies are implemented to improve the efficiency of the designs. IC foundries are adding more design rules to improve the design robustness.

For many high density, high speed DSM designs, power consumption is a major concern. Increasing transistor counts, chip speed, and greater device leakage are driving up both dynamic and static power consumption. For example, increased leakage currents have become a significant contributor to the overall power dissipation of Complimentary Metal Oxide Semiconductor (CMOS) designs beyond

C. Duan et al., *On and Off-Chip Crosstalk Avoidance in VLSI Design*,
DOI 10.1007/978-1-4419-0947-3_1, © Springer Science+Business Media LLC 2010

90 nm. Consequently, the identification, modeling and control of different leakage components is an important requirement, especially for low power applications. The reduction in leakage currents can be achieved using both process and circuit level techniques. At the process level, leakage reduction can be achieved by controlling the device dimensions (channel length, oxide thickness, junction depth, etc) and doping profile of transistors. At the circuit level, leakage control can be achieved by transistor stacking, use of multiple V_T or dynamic V_T devices etc. The high power consumption density also make the heat dissipation critical and often requires advanced packaging.

A critical challenge which we will address in Part I is the performance degradation caused by increased on-chip interconnect delays in advanced processes. As a result of aggressive device scaling, gate delays are reducing rapidly. However, interconnect delays remain unchanged or are reducing at a much slower rate. Therefore the performance of bus based interconnects has become the bottleneck to overall system performance. In many large designs (e.g. SoC, NoC and CMP designs) where long and wide global busses are used, interconnect delays often dominate logic delays. For example, in a 90 nm process, the typical gate delay is ~30 ps. The interconnect delay, however, can easily be a few nanoseconds in a moderate sized chip.

1.2 Overview of On-Chip Crosstalk Avoidance

Once negligible, capacitive crosstalk has become a major determinant of the total power consumption and delay of on-chip busses. Figure 1.1 illustrates a simplified on-chip bus model with crosstalk. In the figure, C_L denotes the *load capacitance*, which includes the receiver gate capacitance and also the parasitic wire-to-substrate parasitic capacitance. C_I is the *inter-wire coupling capacitance* between adjacent signal lines of the bus. In practice, this bus structure is typically modeled as a

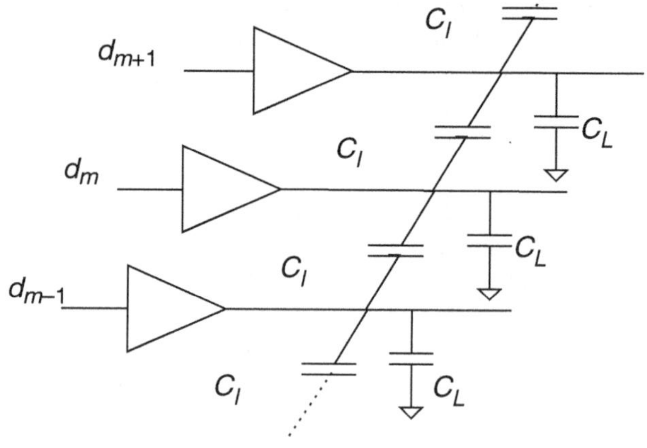

Fig. 1.1 On-chip bus model with crosstalk

distributed RC network, which includes the non-zero resistance of the wire as well (not shown in Fig. 1.1). It has been shown that for DSM processes, C_I is much greater than C_L [61]. Based on the energy consumption and delay models given in [88], the bus energy consumption can be derived as a function of the total crosstalk over the entire bus. The worst case delay, which determines the maximum speed of the bus, is limited by the maximum crosstalk that any wire in the bus incurs. It has been shown that reducing the crosstalk boosts the bus performance significantly [36, 88].

Different approaches have been proposed for crosstalk reduction in the context of bus interconnects. Some schemes focus on reducing the energy consumption, some focus on minimizing the delay and other schemes address both.

The simplest approach to address the inter-wire crosstalk problem is to shield each signal using grounded conductors. Khatri et al. in [61, 62] proposed a layout fabric that alternatively inserts one ground wire and one power wire between every signal wire, i.e., the wires are laid out as ... *VSGSVSGSVS* ..., where *S* denotes a signal wire, *G* denotes a ground wire and *V* denotes a power wire. Any signal wire has a static (V or G) wire on each side, and hence, when it switches, it needs to charge a capacitance of value of $2C_I$. This fabric also enforces a design rule that metal wires on a given layer run perpendicular to wires on layers above or below. The fabric has the advantage of improved predictability in parasitic capacitance and inductance. It automatically provides a low resistance power and ground network as well. Such a fabric results in a decrease in wiring density. Even though the fabric appears to require a large area overhead, experimental results show that on average, the circuit size grows by only 3% if the circuit is implemented as a network of medium-sized programmable logic arrays (PLAs). In the worst case, however, the circuit size can be more than 200% of the conventional layout. Figure 1.2 illustrates such a fabric-based design. Khatri et al. [61] also provides discussions on wire removal in networks of Programmable Logic Arrays (PLA).

The fabric approach is based on passive or static shielding. Active shielding, in contrast, uses active signals to shield the signal of interest. Compared to passive shielding, active shielding is a more aggressive technique that reduces the bus delay by up to 75% [35, 60]. However, the area overhead can be significant. In [60], Kaul et al. examined the concept of active shielding in the context of RLC global interconnects and showed that for RC dominated wires, in-phase switching of shields helps to speed up signal propagation with an acceptable increase in ringing. For wide inductive wires, signal propagation is enhanced and ringing is reduced by

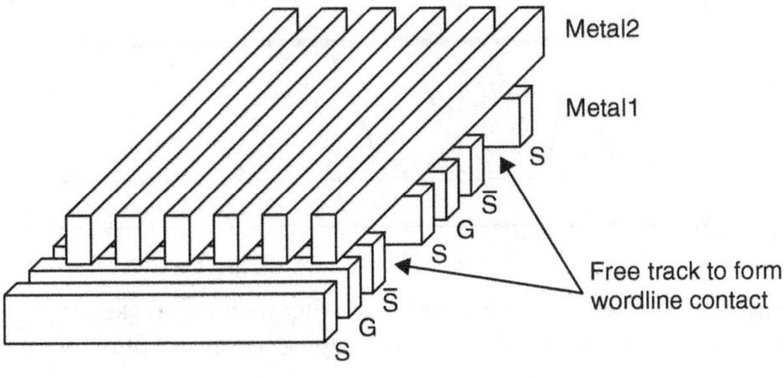

Fig. 1.2 An example of arrangement of conductors in the PLA core [61]

switching the shields in opposite directions. Shields switching in the same phase as the signal have shown up to ~12% and ~33% improvement (compared to passive shields) in delay and transition times respectively for realistic test cases. The opposite phase active shielding concept was used to optimize clock nets and resulted in up to ~40% reduction in transition times. It also resulted in much lower ringing (up to 4.5× reduction). In addition to the huge (>200%) area overhead, active shielding consumes extra power and therefore the performance gains it offers often do not justify these overhead.

Since the worst case crosstalk is a result of simultaneous transition of data bits in synchronous busses, skewing of transition times of adjacent bits can alleviate the crosstalk effect. Hirose and Yasuura reported in [44] that using the technique that intentionally introduces transition time skewing can prevent simultaneous opposite transitions between adjacent wires and consequently reduce the worst case bus delay. This intentional skewing is achieved by adding different time shifts to different bits in the bus. The bus wires alternatively have normal timing and shifted timing, hence no adjacent wires switch at the same time. The shifting in time can be achieved by a delay line or by using a two-phased clock. This technique can be combined with various repeater insertion techniques as well. Figure 1.3 plots the signal transition

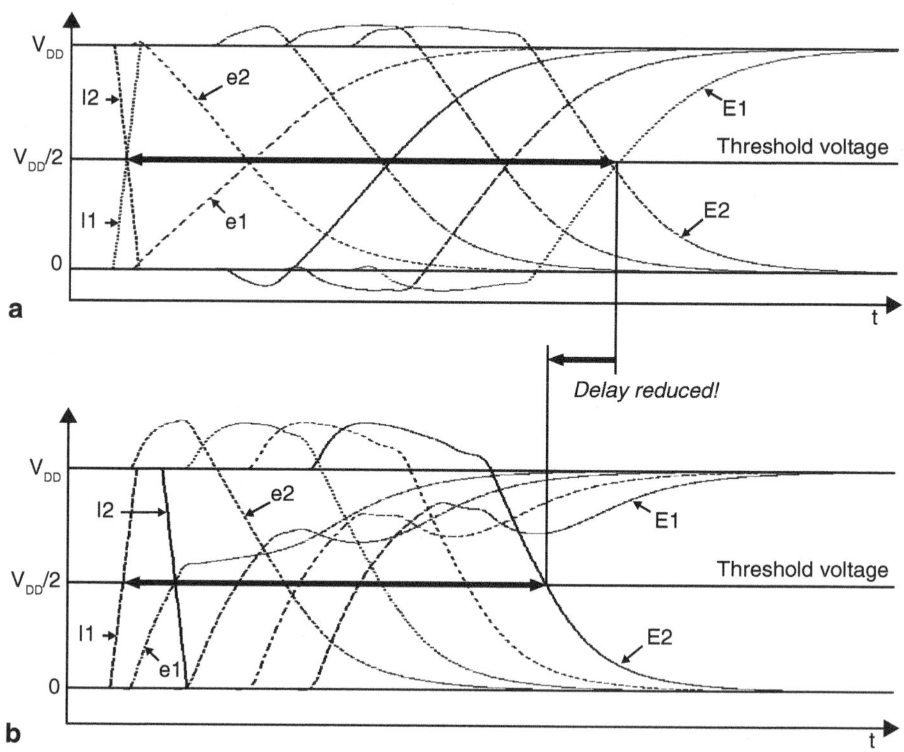

Fig. 1.3 Signal delay improvement using intentional skewing [44]. **a** shows the signal waveform of a synchronous bus and **b** shows the signal waveform of bus with intentional skewing implemented

delay improvement of a time-skewed bus over a conventional bus. Experimental results given in [44] show that the overall bus delay can be reduced by 5–20%. This approach, however, may not improve the maximum bus speed, since it requires a delay overhead between transitions to accommodate the delay shift. Another drawback of such technique is that it requires careful layout tuning. To separate the transitions sufficiently, a large delay may be needed (e.g., a trace of 20 mm requires a delay of ~1 ns). This is rather difficult to implement in most VLSI designs. The delays between the transmitter and receiver also need to be perfectly matched, which is hard to achieve since DSM designs are highly sensitive to process variations. All these constraints limit the employment of such a technique in practice.

The length of the bus plays an important role in bus performances as both the total resistance and the total capacitance are proportional to the trace length. The signal delay is proportional to the product of resistance and capacitance, and therefore grows quadratically with trace length. As a result, the delay of long busses can be significant. One solution is to insert repeaters in long busses. This results a linear increase of delay with trace length and has been adopted widely in design practice. Generally, the repeaters are inserted at the same location for all bits in the bus and are evenly spaced to guarantee the uniform delay. An improved repeater insertion technique called "repeater interleaving" can be used to mitigate the effect of crosstalk. Figure 1.4 illustrates a design based on this technique. Since inverters are commonly used as repeaters (buffers) in a long bus, the direction of the transition in the lth segment of a given line is inverted in the $(l+1)$th segment. Therefore by staggering the repeaters for adjacent bits, the charging and discharging of inter-wire capacitors of the neighboring wires will cancel each others, i.e., the transition on a wire is "pushed" and "pulled" by two adjacent segment of a neighboring wire. Hence the delay increase in one wire segment will be compensated by a delay decrease in the next segment, resulting in each wire charging a capacitance of $2C_I$ on average. Elfadel in [38], Ghoneima and Ismail in [39] investigated the theoretical optimum position for repeater insertions to achieve minimum delay. Again, this technique is a physical design technique and requires careful trace layout and repeater placement.

The average delay can also be reduced using statistical approaches such as data packing and data permutation proposed by Satyanarayana et al. in [86] and variable

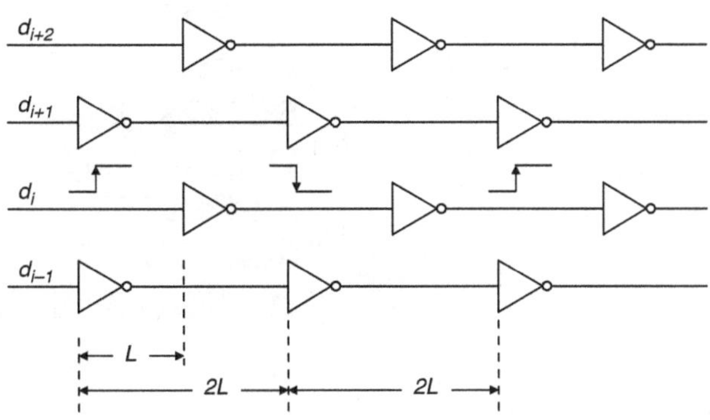

Fig. 1.4 Illustration of repeater interleaving

cycle transmission proposed by Li et al. in [67]. All these techniques exploit the temporal coherency of bus data to minimize the average delay for data transmission. Experimental results show significant improvements in the address bus case, with more than 2.3× speedup reported in [86] and 1.6× speedup reported in [67]. In the case of a data bus, close to 1.5× speedup is reported. These techniques require 2–6 additional wires. However, all statistical approaches require buffering capability on both the transmitter and receiver sides, to handle the non-deterministic delay variation. The worst case delays remain unchanged.

Ho et al. in [45] and Schinkel et al. in [87] proposed using twisted differential pairs for RC limited on-chip busses. A higher than 1 GHz data rate, with single-cycle latency in a 10 mm on-chip wire in a 0.18-μm process was demonstrated in [45]. The bus energy is also reduced duo to reduced signal swing. Schinkel et al. proposed an offset differential structure as shown in Fig. 1.5 and reported a data rate up to 3 GHz [87]. Two major drawbacks of these approaches are the doubling of the number of wires and the higher complexity and power consumption for both the differential transmitter and receiver designs.

Generally, capacitive coupling dominates the delay and power of an on-chip bus. However, on-chip inductive coupling cannot be ignored when the bus speed exceeds a certain limit. He et al. investigated the modeling of inductive crosstalk [42, 43] and they proposed several methods to control inductive crosstalk as well [42]. Their methods focused on a shield insertion and net ordering (SINO) algorithm that takes into account inductive coupling. Several algorithms were proposed to achieve bus designs that are either noise free (NF) or noise bounded (NB).

In addition to research in crosstalk avoidance for bus speed improvement, there are also many publications focus on bus energy consumption reduction. The overall energy consumption of any CMOS circuit is proportional to the total number of transitions on the bus. The instantaneous power consumption on an N-bit bus at cycle i follows the general power dissipation equation by a CMOS circuit [91]

$$P_{bus}(i) = \sum_{j=1}^{N} C_j \cdot V_{dd}^2 \cdot f \cdot p_j(i) \tag{1.1}$$

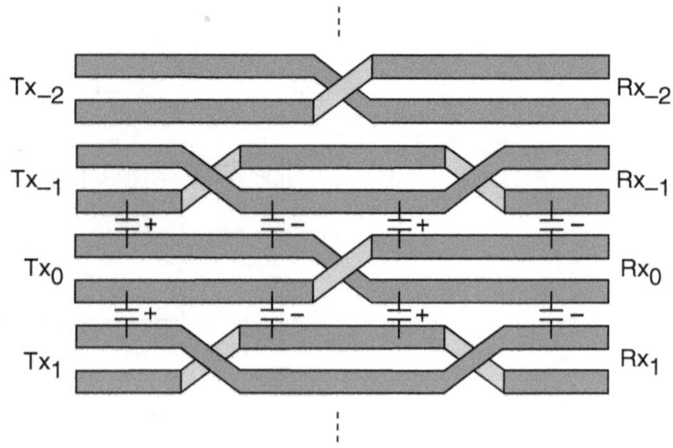

Fig. 1.5 Twisted differential pairs with offset twists [45]

where N is the number of bits in the bus, C_j is the load capacitor on the jth line, V_{dd} is the supply voltage, f the frequency and p_j the activity factor at bit j. Assuming the $C_j = C_{load}$ for all j, the total energy consumption and average power of the bus is

$$P_{bus}(i) = C_{load} \cdot V_{dd}^2 \cdot f \cdot N(i) \tag{1.2}$$

$$\overline{P_{bus}} = C_{load} \cdot V_{dd}^2 \cdot f \cdot \overline{N} \tag{1.3}$$

where $N(i) = \sum p(j)$ is the number of transitions at cycle i and $\overline{N}$ is the average number of transitions per cycle. A reduction in the average power can be achieved not only by reducing C and V_{dd}, but also by lowering $\overline{N}$.

In [65], Kim et al. proposed a coupling driven bus inversion scheme to reduce the transitions on the bus. The basic idea of the scheme is to compute the total number of transitions (including the coupling transitions) at a bus cycle, compare it to a given threshold and invert the bus when the total transition exceeds the threshold. They reported a 30% power saving on average.

In [85], Saraswat, Haghani and Bernard proposed using "T0" or "Gray Code" on busses where data shows pattern regularity. A typical example is an address bus, since the data on an address bus typically changes at a regular increment. Neither coding schemes is suitable for general purpose busses with random data since the reduction in power in these schemes is highly dependent on the coherency of the data.

The effect of crosstalk on bus speed is not limited to on-chip busses. Crosstalk also impacts off-chip bus performance. In addition to the differences in wire dimensions, a major difference between on-chip and off-chip busses is that inductive coupling plays a more significant role in the overall crosstalk for off-chip busses. Part II presents a comprehensive discussion on off-chip bus crosstalk.

1.3 Bus Encoding for Crosstalk Avoidance

The majority of Part I is devoted to bus encoding related topics for on-chip bus crosstalk avoidance. Crosstalk avoidance bus encoding techniques manipulate the input data before transmitting them on the bus. Bus encoding can eliminate certain undesirable data patterns and thereby reduce or eliminate crosstalk, with much lower area overhead than the aforementioned straightforward shielding techniques [36, 88, 90, 99, 100]. These types of codes are referred to as *crosstalk avoidance codes* (CACs) or *self shielding codes*. Depends on their memory requirements, CACs can be further divided into two categories: memoryless codes and memory-based codes. Memory-based coding approaches generate a codeword based on the previously transmitted code and the current data word to be transmitted [34, 100]. On the receiver side, data is recovered based on the received codewords from the current and previous cycles. The memoryless coding approaches use a fixed code book to generate codewords to transmit. The codeword is solely dependent on the input data. The decoder in the receiver uses the current received codeword as the only input in order to recover the data.

Among all the memoryless CACs proposed, two types of codes have been heavily studied. The first is called *forbidden pattern free* (FPF) code and the second type of code has the property that between any two adjacent wires in the bus, there will be no transition in opposite directions in the same clock cycle. Different names have been used in the literatures for the second type of codes. In this paper, we refer these codes as *forbidden transition free* (FTF) CACs. Both FPF-CACs and FTF-CACs yield the same degree of delay reduction as passive shielding while requiring much less area overhead. Theoretically, the FPF-CAC has slightly better overhead performance that the FTF-CAC. In practice, for large size bus, this difference is negligible.

In this monograph, we also discuss some memory-based encoding techniques. Two such coding approaches are offered in this monograph: the "code pruning" and the "ROBDD-based" method. Our theoretical analysis shows that memory-based codes offer improved overhead efficiency, compared to memoryless codes, and the experimental results confirm this analysis.

Crosstalk avoidance codes cannot be put in practical use without efficient encoder and decoder (CODEC) implementations. We dedicate a portion of this monograph to discussions of CODEC designs for CACs. We look at the impact of mapping schemes on the CODEC complexity. We present several CODEC designs that are highly efficient compared to CODECs designs using conventional approaches. The "group complement" CODEC design uses bus partitioning to ensure low CODEC complexity and high speed. Two CODEC designs based on the Fibonacci numeral system (FNS) achieve efficient area overhead while offering low complexity and high speed implementations.

Even though most of the codes and CODECs discussed in this monograph are designs for binary busses, we also investigate some crosstalk avoidance schemes for multi-valued busses and develop several encoding schemes for multi-valued busses that allows very low power and high speed bus interconnect. Study shows that the multi-valued busses are advantageous over the binary busses in many cases. The monograph also addresses the the driver and receiver designs as they are often more challenging than the binary drivers/receivers.

1.4 Part I Organization

The remainder of Part I is organized as follows: Chap. 2 provides physical and mathematical models for bus signal delay and energy consumption, which account for capacitive coupling between wires. Using these models, we establish a bus crosstalk classification system. This classification system serves as the foundation for the subsequent chapters. In Chap. 3, we discuss memoryless CACs, present design techniques/algorithms to construct these codes, and provide the analysis and experimental results to compare their area overhead. Chapter 4 is dedicated to CODEC designs for memoryless CACs. Several CODEC design techniques are presented including "group complement", a design technique based on bus partitioning, and several "Fibonacci Numeral System"-based designs. Through analysis and experiments we show that these CODEC designs are efficient in terms of area overhead, complexity and speed, compared to conventional schemes. In Chap. 5, we study memory-based codes and show that these codes have better area overhead than memoryless approaches. Several code design algorithms are given in the chapter along with experimental results to validate our analysis. Chapter 6 extends the idea of crosstalk avoidance encoding to multi-valued busses. We first generalize the bus classification system to the multi-valued bus domain. We then present a bit-to-bit mapping scheme and two types of codes specifically designed for ternary busses. Experimental results are given to show that both coding schemes offer high speed and energy efficiency. We also discuss the challenges in bus driver and receiver circuit designs in this chapter. Chapter 7 summarizes the work present in part I and offers some insight on the directions for future work.

Chapter 2
Preliminaries to On-Chip Crosstalk

This chapter provides background information and notation that will be needed for subsequent chapters. Section 2.1 focuses on modeling of on-chip bus parasitics. We show the impact of wire dimensions and bus geometry on resistance and capacitance, and describe the equivalent circuit for the on-chip bus. We then derive the mathematical expression for both the signal delay and the energy consumption of a bus based on our models. In Sect. 2.2 we establish a bus classification system. This bus classification system serves as the foundation for the rest of this thesis.

2.1 Modeling of On-Chip Interconnects

Complementary metal-oxide-semiconductor (CMOS) on Silicon is the most commonly used process for digital circuit design. Other processes include Silicon bipolar, GaAs, SiGe, Bi-CMOS or Silicon on insulator (SoI). The models built in this work are based on CMOS process. However, all techniques discussed in this part are technology independent, with minor differences arising from circuit implementations.

The circuitry of a digital VLSI design is an assembly of gates and interconnects. Gates are the active circuits that implement logic operations. Interconnects are simply conductors that connect the logic gates. Aluminum or copper are commonly used to implement interconnect. Interconnects can be separated into *local interconnects* and *global interconnects*. Local interconnects generally refer to connections between gates within functional block. Global interconnects provide connections between functional blocks. The length of local interconnects is relatively short. Global interconnects, on the other hand, can traverse the entire chip and be fairly long.

Interconnects introduce additional delay to the signals that travel through them [41, 68]. The delays of local interconnect can sometimes be significant but they are relatively easier to deal with through physical layout, floorplanning, circuit re-timing or logic optimization.

Delays in global interconnect present a much greater challenge in high-performance designs. Because of their longer wire lengths (which can be as high as ten's of millimeters), global interconnect delays are much larger than gate delays and local

C. Duan et al., *On and Off-Chip Crosstalk Avoidance in VLSI Design*,
DOI 10.1007/978-1-4419-0947-3_2, © Springer Science+Business Media LLC 2010

Fig. 2.1 3-D structure of an on-chip conductor

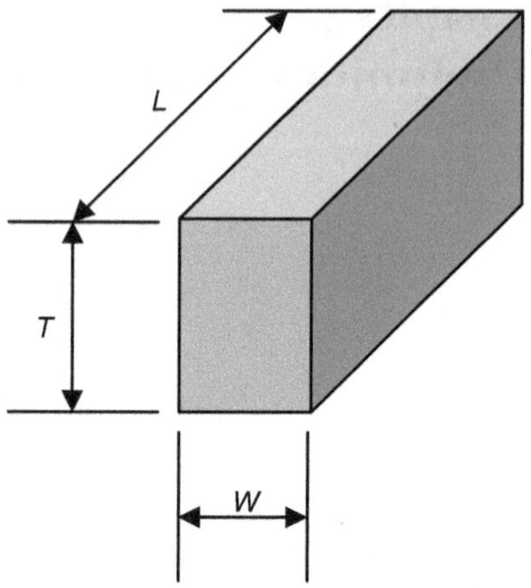

interconnect delays. Global interconnects also contribute a large portion to the overall power consumption. Lin pointed out in [41] that the total length of interconnects on a chip can reach several meters and as much as 90% of the total delay in DSM designs can be attributed to global interconnects.

Figure 2.1 shows the 3-dimensional structure of a conductor (wire) that is used in VLSI interconnect. The length of the wire, L, is determined by the distance between the components it connects and can be controlled by designers during the physical layout. Even so, in large and complex designs, it may not be feasible to place all components close to each other. The width of the wire, W, can also be adjusted during physical layout. However, the minimum width, W_{min}, is determined by the fabrication process. The height T is a constant for a given fabrication process and cannot be modified by the IC designers.

Even though other types of busses such as serial or asynchronous busses can be used for on-chip interconnect, the majority of the digital VLSI circuits employ synchronous parallel busses due to their simplicity, good timing predicability and high throughput. A parallel global bus is typically implemented as a group of wires routed in parallel as shown in Fig. 2.2. The lengths of the wires are approximately matched. In practice, upper metal layers are used for global interconnects. In addition to W, T and L of individual wires, two more parameters are required to specify the bus: the spacing between wires, P, and the distance between the bus wires and the substrate, H. Similar to W_{min}, the minimum spacing P_{min} is determined by the fabrication process and generally it can be assumed that $P_{min} = W_{min}$. In general, P_{min} and W_{min} are used whenever feasible to minimize the total area a bus occupies. For a given metal layer, H is also determined by the fabrication process.

For on-chip busses, two important electrical parameters of the wires of a bus are the resistance R and the parasitic capacitance C. Their values are directly related to the bus geometry. The resistance of a conductor is given as $R = \frac{\rho L}{WT}$, where ρ is the resistivity of the material. Since ρ is determined by the material used for interconnects (Al or Cu), it is deemed fixed. To maintain low resistance of the conductors, we can

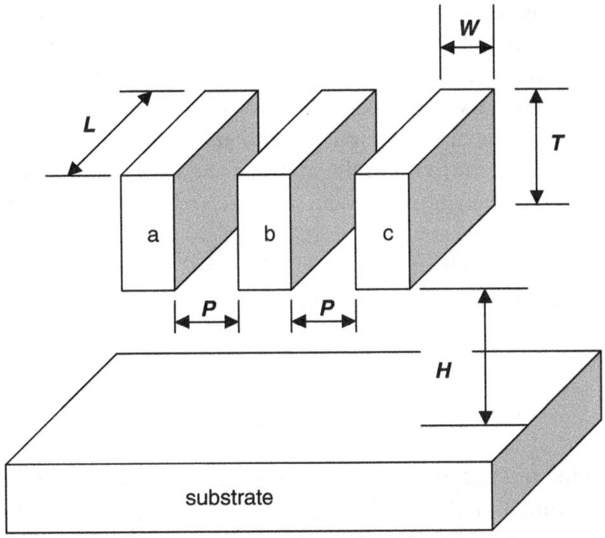

Fig. 2.2 A simplified 3-D structure of an on-chip bus

choose materials with high conductivity, or vary bus geometry and wire dimensions including W, L, and T. An increase in T and W results in lower resistance, so does decreasing L. However, as discussed previously, a small L can be attained through floorplanning in the layout stage, but is not always feasible in complex designs. Also, T is determined by the fabrication process and is not under designer's control. Finally, increasing W increases the die area occupied by the bus.

The modeling of parasitic capacitance is less straightforward. There have been numerous publications on accurate modeling and extraction of interconnect capacitance using advanced methods such as finite difference, finite element, boundary element etc. [14, 25, 28, 94]. Interested readers can find a comprehensive overview of interconnect capacitance modeling in [14]. We will only give some simplified models here.

Any two conductive surfaces (such as two wires in an on-chip bus, or a wire and the grounded substrate) form a capacitor. The capacitance is a function of the geometries of the conductors. Two models that can be used to determine the capacitor between two parallel conductors: the parallel-plate capacitor model and the fringe capacitor model.

For a given wire in a bus, two dominant parasitic capacitance are the substrate capacitance and inter-wire capacitance. The substrate capacitor is formed between a wire and the substrate and is denoted by C_L. The inter-wire capacitance is a capacitor between two adjacent wires routed in parallel and is denoted by C_I. There are other parasitic capacitors beside C_L and C_I and in most cases they are negligible [14]. For example, the *overlap capacitance* refers to the capacitance between wires on different layers. In practice, wires on two adjacent layers are routed orthogonally (i.e., if on the k-th layer signals are routed North-South, then traces on the $(k \pm 1)$-th layers are routed East-West.). Also the coupling capacitance between two coplanar wires is negligible if these two wires are not adjacent to each other (because wires in between act as shields which terminate the electrical field).

Using the model given in Fig. 2.2, it has been shown in [61] that to estimate the capacitance between adjacent wires, the parallel-plate model is applicable when $T \gg P$, where the capacitance can be expressed as $C_I = k\epsilon_0 \frac{T \cdot L}{P}$. Here k is the relative permittivity of the material and ϵ_0 is the permittivity of free space. If $T/P \sim 1$ or less, the fringing model applies and the capacitance is given as $C_I \propto \log(P)$. Similarly, to estimate the capacitance of a wire to the substrate, the parallel-plate model is applicable if $W \gg H$, and $C_L = k\epsilon_0 \frac{W \cdot L}{H}$. If $W/H \sim 1$, then the fringing model applies and $C_L \propto \log(H)$.

For convenience of the ensuring discussion, we first define

$$\lambda = \frac{C_I}{C_L}. \tag{2.1}$$

as the ratio of the *inter-wire capacitance* to the *substrate capacitance*. Once again λ is a function of the bus geometry. For processes with feature size of 1 μm or greater, $\lambda < 1$, indicating that C_I is negligible compared to the total capacitance. However, in DSM processes, $\lambda \gg 1$ [61] and C_I contributes a dominant fraction of the total capacitance. Table 2.1 shows the technology evolution and the changes in the circuit geometry. Table 2.2 lists the parasitic capacitance for the first 6 metal layers in 3 different technologies. The parameters were extracted using a 3-D extraction tool space3D [6] and based on a "strawman" model. The original work was conducted by Khatri in [61] and we were able to validate all the values independently.

Table 2.1 DSM process parameters [68]

Technology (μm)	0.25	0.18	0.15	0.13	0.09
Vdd (V)	2.5	1.8	1.5	1.2	1
Core gate oxide (Å)	50	32	26	20	16
No. of metal layers	5	6	7	9	10
Dielectric constant	4.1	3.7	3.6	2.9	2.9
Metal pitch–M1(μm)	0.64	0.46	0.39	0.34	0.24
Metal pitch–Others (μm)	0.8	0.56	0.48	0.41	0.28
Resistivity–M1 (Ω−cm)	75	80	120	90	100
Resistivity–N+/Poly (Ω/Ct)	7	8	13	11	20
Resistivity–via (Ω/Ct)	4	7	9	1	1.5
Ca (aF/μm)	15	11	11	15	14
Cf (aF/μm)	10	8	8	11	9
Cc (aF/μm)	80	90	97	105	94
NMOS Vt (V)	0.53	0.42	0.42	0.28	0.27
PMOS Vt (V)	0.53	0.5	0.42	0.42	0.32
NMOS Idsat (μA/μm)	600	600	580	450	640
PMOS Idsat(μA/μm)	270	270	260	270	280
NMOS Ioff (pA/μm)	10	20	35	320	500
PMOS Ioff (pA/μm)	1.5	2.5	40	160	500
kgates/mm^2	58	100	120	221	400
Gate delay (ps)	40	32	20	17	12

Table 2.2 Parasitic capacitance (in $pF/\mu m$) at different metal layers in three different processes. All values were extracted using SPACE3D [61]

Technology	0.25 µm		0.10 µm		0.05 µm	
Layers	C_L	C_I	C_L	C_I	C_L	C_I
1	10.96	40.76	9.97	49.35	15.75	46.82
2	0.86	27.79	0.92	48.05	1.64	47.58
3	1.77	40.92	2.58	45.84	5.38	48.28
4	0.96	41.83	1.10	44.08	1.2	46.66
5	2.02	23.36	2.31	39.84	3.90	39.15
6	10.27	33.48	1.07	40.59	1.51	38.33

Two equivalent circuits for data busses with parasitic resistance, R_W, and parasitic capacitance, C_W, included are given in Fig. 2.3. The driver resistance is assumed to be R_D. In Fig. 2.3a, the bus is modeled as a simple lumped RC network. Such a model is accurate at low data rates and for moderate trace length. In general this model is accurate if the driver rise or fall times (τ) is such that $\tau > R_W C_W$. As the data rate goes up and the trace length becomes longer, the lumped RC model becomes inadequate and the bus needs to be modeled using a more sophisticated

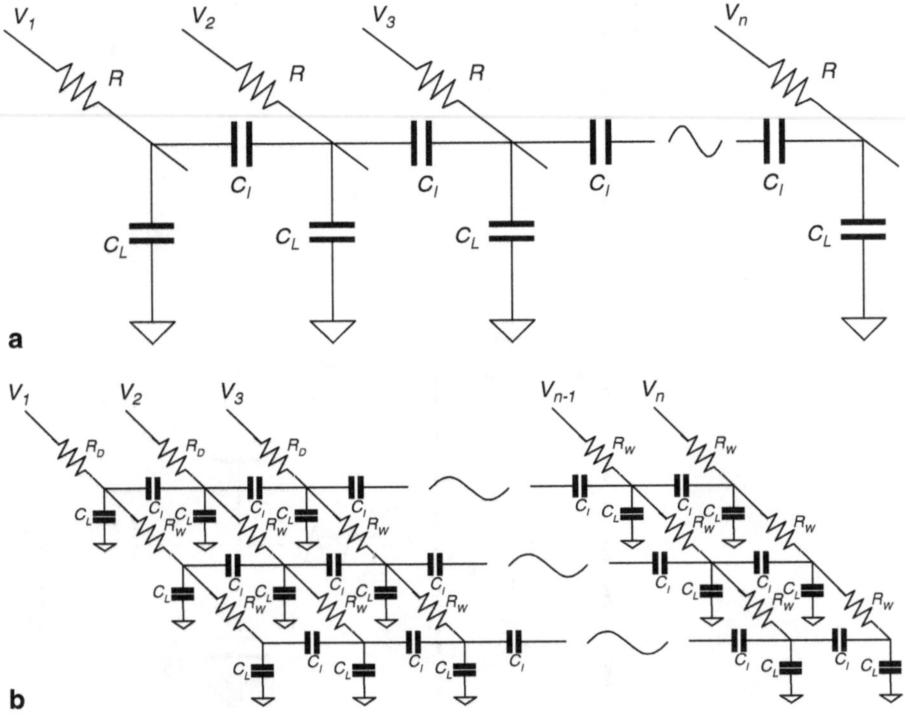

Fig. 2.3 Equivalent circuit for the data bus **a** Lumped RC model **b** Distributed RC model using 2π-sections

distributed RC network as shown in Fig. 2.3b. The condition for which a distributed network is required is $\tau < R_W C_W$.

Accurate driver circuit modeling is also necessary in characterizing the bus behavior. For now, we assume that busses for on-chip interconnects are binary and therefore the driver only outputs two voltages representing logic high or logic low. The steady state voltage for logic high can be approximated as V_{dd} and that for logic low as 0V. This assumption holds for most digital CMOS circuits.

In practice, bus drivers are typically implemented using inverters that are properly sized to meet the speed requirements of the bus. The maximum driver current is approximately proportional to the gate width of the transistors of the driving inverter. Figure 2.4 shows the equivalent circuits of a driver in the output-high and the output-low states respectively. The driver is approximated as a resistor connected to V_{dd} or

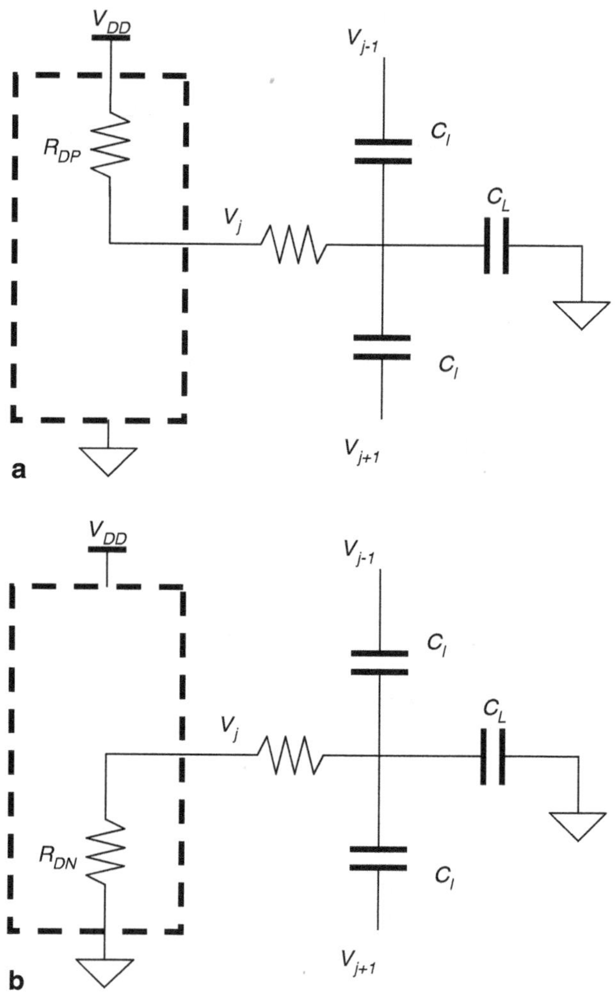

Fig. 2.4 Equivalent circuits for the drivers **a** V_j high **b** V_j low

0V depends on the output logic state. In the output high state, the drive charges the capacitors on the wire through a resistor with a resistance of R_{DP} until the voltage on the line reaches V_{dd}. During the output-low, the capacitors on the bus wire is discharged through a resistor with resistance of R_{DN}. The value of R_{DP} and R_{DN} are determined by the transistor sizes of the driver and it can be assumed that the drivers are sized so that $R_{DP} = R_{DN} = R_D$. We also assume the all drivers in the same bus have the same sizes. In other words, all drivers of the same bus have a driver resistance of R_D.

Using the equivalent circuits given in Figs. 2.3 and 2.4, we can derive the bus energy consumption expression. Let $E_{r,j}$ be the energy needed to drive the jth line in the bus from logic low to logic high, i.e. to charge all the capacitors on the jth line from 0V to V_{dd}. Let $E_{f,j}$ be the energy needed to drive the jth line from logic low to logic high, i.e. to discharge the capacitors from V_{dd} to 0V. Since no current flows through V_{dd} in output-high state, we know

$$E_{f,j} \equiv 0. \tag{2.2}$$

To compute the energy consumed to charge the bus in the output-high state, $E_{r,j}$, we know

$$E_{r,j} = \int_{t^-}^{t^+} V_{dd} \cdot I_j(t)dt \tag{2.3}$$

where t^- denotes the time before the transition, such that $V_j(t^-) = 0$ and t^+ denotes the time after the transition when the steady state voltage is reached, i.e., $V_j(t^+) = V_{dd}$. $V_j(t)$ is simplified as V_j hereafter. Since

$$I_j(t) = C_L\frac{dV_j}{dt} + C_I\frac{dV_{j,j-1}}{dt} + C_I\frac{dV_{j,j+1}}{dt} \tag{2.4}$$

where $dV_{j,k} = dV_j - dV_k$. Now Eq. 2.3 can be rewritten as

$$\begin{aligned}
E_{r,j} &= \int_{t^-}^{t^+} V_{dd} \cdot \left(C_L\frac{dV_j}{dt} + C_I\frac{dV_{j,j-1}}{dt} + C_I\frac{dV_{j,j+1}}{dt}\right)dt \\
&= C_L \cdot V_{dd}\int_{t^-}^{t^+} dV_j + C_I \cdot V_{dd}\int_{t^-}^{t^+} dV_{j,j-1} + C_I \cdot V_{dd}\int_{t^-}^{t^+} dV_{j,j+1} \\
&= C_L \cdot V_{dd}(\Delta V_j + \lambda\Delta V_{j,j-1} + \lambda\Delta V_{j,j+1}) \\
&= (1 + 2\lambda - \lambda\delta_{j,j-1} - \lambda\delta_{j,j+1})C_L \cdot V_{dd} \cdot V_j
\end{aligned} \tag{2.5}$$

where

$$\Delta V_j = V_j(t^+) - V_j(t^-), \tag{2.6}$$

$$\Delta V_{j,k} = \Delta V_j - \Delta V_k, \tag{2.7}$$

$$\delta_{j,k} = sgn(\Delta V_j)\frac{\Delta V_k}{V_{dd}}, \tag{2.8}$$

$\delta_{j,k}$ is the *normalized relative voltage change* of the kth line with respect to the jth line, and $\delta_{j,k} \in \{0, \pm 1\}$.

Equations 2.2 and 2.5 can be combined into a single expression

$$\begin{aligned} E_j &= E_{r,j} + E_{f,j} \\ &= (1 + 2\lambda - \lambda\delta_{j,j-1} - \lambda\delta_{j,j+1}) C_L \cdot V_{dd}^2 \cdot V_j \\ &= E_j^L + E_j^I \end{aligned} \tag{2.9}$$

where $E_j^L = C_L \cdot V_{dd} \cdot V_j$ and $E_j^I = (2 - \delta_{j,j-1} - \delta_{j,j+1})\lambda C_L \cdot V_{dd} \cdot V_j$. Equation 2.9 can be interpreted as the total energy to drive a line and it can be expressed as the sum of the energy to charge the substrate capacitor, E_j^L, and the energy to charge the inter-wire capacitors, E_j^I. Equation 2.9 also shows that the energy consumption is independent of R_D and therefore the driver size.

For a bus, the total energy consumption is simply the summation of energy on each individual line, and is expressed as follows

$$E_{total} = \sum_{j=1}^{N} E_j = \sum_{j=1}^{N} E_j^L + \sum_{j=1}^{N} E_j^I. \tag{2.10}$$

Again, the total energy can be separated into two portions: the energy to charge the load capacitors and the energy to charge the inter-wire capacitors.

To derive the signal delay for a bus with crosstalk, we first recall the derivation of the delay of a single wire when no inter-wire crosstalk is present. It is derived by first noting that the current flowing into C_L and R is the same. Therefore in the output high state, the current is

$$I_j(t) = C_L \frac{dV(t)}{dt} = \frac{V_{dd} - V(t)}{R} \tag{2.11}$$

It can be rewritten as

$$R \cdot C_L \cdot dV(t) = (V_{dd} - V(t))dt \tag{2.12}$$

Integration on both sides of Eq. 2.12, we get

$$\begin{aligned} R \cdot C_L \int_0^{V(\tau)} dV(t) &= \int_0^{\tau} (V_{dd} - V(t))dt \\ R \cdot C_L \cdot V(\tau) &= V_{dd} \cdot \tau - \int_0^{\tau} V(t)dt \end{aligned} \tag{2.13}$$

The solution to the differential equation Eq. 2.13 is

$$V(\tau) = V_{dd}(1 - e^{-\frac{\tau}{RC}}) \tag{2.14}$$

with the initial condition $V_j(0) = 0V$, the rise time to reach $V_j(t) = \alpha V_{dd}$ is

$$\tau(\alpha) = -RC \ln(1 - \alpha) \tag{2.15}$$

$\tau(0.5) = 0.69RC$ gives the rise time for the signal to reach the switching point (which is typically $0.5V_{dd}$ for CMOS) and is generally used for the delay calculation. Other rise time values of interest are $\tau(0.8) = 1.6RC$ and $\tau(0.9) = 2.3RC$.

Based on the previous assumption that the drivers are properly sized to satisfy $R_D = R_U = R$, we can easily find out that the fall time for a logic high to logic low transition is the same as the rise times.

$$V(\tau) = V_{dd}e^{-\frac{\tau}{RC}} \tag{2.16}$$

$$\tau(\alpha) = -RC\ln(\alpha) \tag{2.17}$$

When inter-wire capacitive crosstalk is included, Eqs. 2.11 and 2.12 become

$$C_L\frac{dV_j(t)}{dt} + C_I\frac{dV_{j,j-1}(t)}{dt} + C_I\frac{dV_{j,j+1}(t)}{dt} = \frac{V_{dd} - V_j(t)}{R} \tag{2.18}$$

$$RC_L \cdot dV_j(t) + RC_I \cdot dV_{j,j-1}(t) + RC_I \cdot dV_{j,j+1}(t) = (V_{dd} - V_j(t))dt \tag{2.19}$$

Rewrite the left hand side of Eq. 2.19 as

$$RC_L[dV_j(t) + \lambda(2 \cdot dV_j(t) - V_{j-1}(t) - dV_{j+1}(t))]$$

and perform integration on both sides, it becomes

$$RC_L\int[dV_j(t) + \lambda(2 \cdot dV_j(t) - V_{j-1}(t) - dV_{j+1}(t))] = \int[V_{dd} - V_j(t)]dt \tag{2.20}$$

To solve Eq. 2.20, $dV_{j-1}(t)$ an $dV_{j+1}(t)$ need to be known. Since all drivers and wires in a bus are identical, we can assume $dV_{j+1}(t) = \delta_{j,j+1}dV_j(t)$ and $dV_{j-1}(t) = \delta_{j,j-1}dV_j(t)$, where $\delta_{j,k} \in \{0, \pm 1\}$. Here Eq. 2.20 can be simplified to

$$RC_L\int(1 + 2\lambda - \lambda\delta_{j,j-1} - \lambda\delta_{j,j+1})dV_j(t) = \int[V_{dd} - V_j(t)]dt. \tag{2.21}$$

and the rise time can be derived in similar fashion as

$$\tau_j(\alpha) = -RC_L(1 + 2\lambda - \lambda\delta_{j,j-1} - \lambda\delta_{j,j+1})\ln(1-\alpha), \tag{2.22}$$

and the 50% switch point delay is

$$\tau_j(\alpha) = -0.69RC_L(1 + 2\lambda - \lambda\delta_{j,j-1} - \lambda\delta_{j,j+1}). \tag{2.23}$$

The fall time can be derived in similar fashion.

2.2 Crosstalk Based Bus Classification

We define the *effective capacitance* of the jth line in a bus as

Definition 2.1

$$C_{eff,j} = C_L \cdot (1 + \lambda(2 - \delta_{j,j-1} - \delta_{j,j+1})). \tag{2.24}$$

Equations 2.22, 2.9 and 2.10 can all be expressed as functions of $C_{eff,j}$:

$$\tau_j(\alpha) = k \cdot V_{dd} \cdot C_{eff,j} \tag{2.25}$$

$$E_j = C_{eff,j} \cdot V_{dd} \cdot V_j \tag{2.26}$$

$$E_{total} = \sum_{j=1}^{n} C_{eff,j} \cdot V_{dd} \cdot V_j \tag{2.27}$$

Equation 2.25 shows that the delay of the jth line is linear in $C_{eff,j}$. Since the value of $C_{eff,j}$ is affected by the transitions on the jth, the $(j-1)$th and the $(j+1)$th lines (i.e, on the wire of interest as well as its immediate neighbors on both sides), the delay is affected by transition patterns on the bus signals $b_{j-1}b_jb_{j+1}$, where b_j is the jth signal. We can classify the 3-bit transition patterns based on the value of $C_{eff,j}$. Under the assumption of $\lambda \gg 1$, we can ignore the C_L term and the possible values of $C_{eff,j}$ are therefore $0 \cdot C_I$, $1 \cdot C_I$, $2 \cdot C_I$, $3 \cdot C_I$ and $4 \cdot C_I$. The corresponding transition patterns are referred to as *0C, 1C, 2C, 3C* and *4C* transition patterns as shown in Table 2.3. The last column in Table 2.3 gives example transition patterns on the three adjacent bits of the bus, $b_{j-1}b_jb_{j+1}$.

If we rewrite Eq. 2.24 as

$$\begin{aligned} C_{eff,j} &= C_L \cdot [1 + \lambda((1 - \delta_{j,j-1}) + (1 - \delta_{j,j+1})] \\ &= C_L + C_{lw,j} + C_{rw,j}, \end{aligned} \tag{2.28}$$

it separates $C_{eff,j}$ into three components: the intrinsic capacitance C_L, the crosstalk capacitance to the wire on the left side, $C_{lw,j} = \lambda(1-\delta_{j,j-1})C_L$, and the capacitance to the wire on the right side, $C_{rw,j} = \lambda(1-\delta_{j,j+1})C_L$. It is easy to see that $C_{lw,j}, C_{rw,j} \in \{0, 1C_I, 2C_I\}$.

From Eqs. 2.24 or 2.28, we can observe that the minimum value $\min(C_{eff}) = C_L$ occur when $\delta_{j,j-1} = 1$ and $\delta_{j,j+1} = 1$, corresponding to the $0C$ transition patterns, the

Table 2.3 Transition pattern classification based on crosstalk

Crosstalk class	C_{eff}	Sample transition patterns
0C	C_L	$000 \rightarrow 111$
1C	$C_L(1+\lambda)$	$011 \rightarrow 000$
2C	$C_L(1+2\lambda)$	$010 \rightarrow 000$
3C	$C_L(1+3\lambda)$	$010 \rightarrow 100$
4C	$C_L(1+4\lambda)$	$010 \rightarrow 101$

maximum values $\max(C_{eff}) = (1+4\lambda)C_L$ occurs when $\delta_{j,j-1} = -1$ and $\delta_{j,j+1} = -1$, corresponding to the $4C$ patterns. Given that the signal delay τ is linear in $C_{eff,j}$, the classification system in Table 2.3 is also a classification system for the speed of individual bus wires.

For a parallel bus, the speed is determined by the worst case delay over all the bits in the bus. We can similarly classify the bus based on *the worst case delay on any bit of the bus*. As an example, suppose that 1101, 0010 and 1100 are transmitted on a 4-bit bus ($b_1b_2b_3b_4$) in cycle k, $k+1$ and $k+2$. The worst case delay is $4C$ on b_3 during the transition from cycle k to $k+1$. Such a bus is therefore classified as a $4C$ bus.

A bus that transmits random data will experience $\max_j\{C_{eff,j}\} = (1+4\lambda)C_L$ and therefore the bus speed must be specified from Eq. 2.25 to accommodate the worst case crosstalk delay, $k \cdot V_{dd} \cdot C_L(1+4\lambda)$. We refer to this as τ_{4C}. In general, we refer to the $k \cdot C$ crosstalk delay as τ_{kC}. Based on Table 2.3 and Eq. 2.25, we observe that when $\lambda \gg 1$, by eliminating $4C$ crosstalk on ALL lines in the bus, we can increase the maximum speed of the bus by ~33%. If we can eliminate $3C$ AND $4C$ crosstalk on all lines, the bus can be sped up by ~100%.

The analysis given above can be easily verified using circuit-level simulation. We implemented several bus circuits in SPICE [76] and verified the bus delay via circuit level simulations. The bus circuit was built using a BSIM4 [2] 0.1 μm process models. The traces are assumed to be routed on Metal4. All parasitic values were obtained by a 3D parasitic extraction tool, SPACE3D [6] (and they match the data reported in [61]). The distributed RC transmission line model shown in Fig. 2.3b was used. Table 2.4 reports the worst-case delays of bus signals. Column 1 lists the lengths of the bus wires. Column 2 reports the driver sizes in multiples of a minimum-sized driver (a minimum-sized driver in this case consists of a PMOS device with channel width of 0.3 μm and a NMOS device with channel width of 0.1 μm. Columns 3–7 report the worst case delay of the bus in picoseconds, assuming that no greater than $0 \cdot C$ through $4 \cdot C$ crosstalk patterns are allowed, respectively. A sample SPICE plot for the delays of signals representing each of the 5 classes of crosstalk is shown in Fig. 2.5.

The delay values in Table 2.4 and Fig. 2.5 show that a bus which has $3C$ and $4C$ pattern removed (e.g., by passive shielding) would reduce its worst case delay by

Table 2.4 Delay comparison (in ps) for different driver sizes and trace lengths

Trace length (mm)	Buffer size	0c	1c	2c	3c	4c
5	30×	83	121	241	516	665
5	60×	108	131	213	399	402
5	120×	96	117	136	196	279
10	30×	102	153	437	912	1026
10	60×	131	164	413	722	919
10	120×	114	137	270	379	548
20	30×	153	203	793	1068	1586
20	60×	164	206	691	1161	1561
20	120×	134	177	580	969	1365

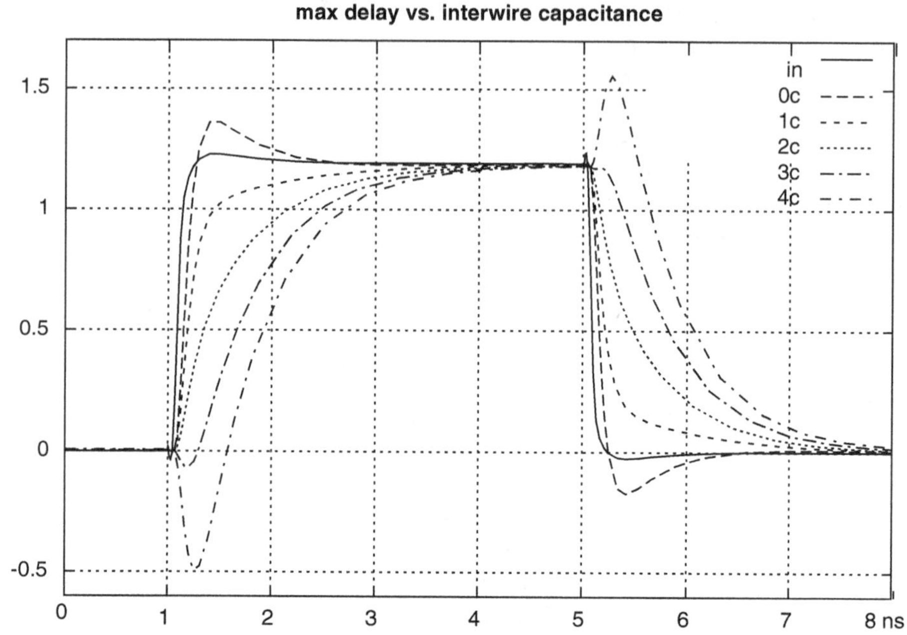

Fig. 2.5 Sample signal waveforms under different crosstalk conditions (60× driver, 2 cm trace length, 0.1 μ process)

at least a factor of 2, regardless of bus trace length and driver strength. This results in significant savings for longer buses. Considering a 2 cm bus driven by a 60× minimum driver, the worst case delay would drop from 900 to 400 ps. All these results confirm our analysis that a bus can be sped up by avoiding certain undesirable crosstalk classes. This is the foundation of the *crosstalk avoidance bus encoding*.

2.3 Bus Encoding for Crosstalk Avoidance

The basic operation of on-chip bus encoding for crosstalk avoidance is illustrated in Fig. 2.6. An n-bit input vector $V_{unc} = b_1 b_2 \cdots b_n$ is first encoded into an m-bit encoded vector, $V_{enc} = d_1 d_2 \cdots d_m$, which is transmitted over the bus. The received vector is decoded to recover the original data $b_1 b_2 \cdots b_n$. The encoded data consists of specially designed codewords which guarantee that some high classes of crosstalk transition patterns never occur. By eliminating these classes of crosstalk

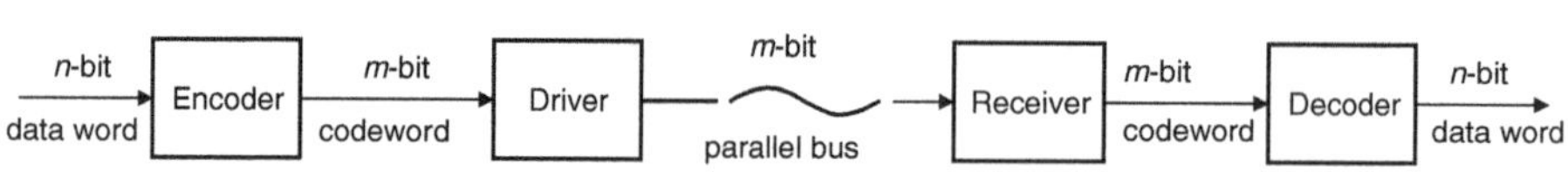

Fig. 2.6 Bus encoding for crosstalk avoidance

patterns, the bus can be operated at a higher speed, while consuming lower power. Such techniques to improve bus speed/power are referred to as *Crosstalk Avoidance Encoding* techniques, and the encoded data are referred to as *Crosstalk Avoidance Codes (CACs).*

Compared to an uncoded bus, a crosstalk avoidance encoded bus requires an encoder on the transmitter side and a decoder on the receiver side. The original data is the input to the encoder and is referred to as the *data word*, while the output of the encoder is referred to as the *codeword* in this thesis. The input to the decoder is the codeword, and its output is the recovered data word.

Typically, the number of bits of the codewords is higher than that of the datawords, i.e., $m > n$. This is not always the case, however, as we will show in Chap. 6. The additional wires used for a coded bus are considered as *area overhead* because of the additional routing space these wires occupy. The overhead percentage is defined as in Eq. 2.29. Note that the area of the encoder and decoder (CODEC) circuit are not included in the calculation of the area overhead. Instead, the encoder and decoder size are evaluated separately. This allows us to evaluate the code efficiency and CODEC efficiency separately.

$$OH(n) = \frac{m-n}{n} \tag{2.29}$$

In this monograph, we use the following set of rules as unified criteria to evaluate the the performance of different CACs:

- Bus speedup: The ratio of the maximum speed of the coded bus to the maximum speed of the uncoded bus.
- Area overhead: The ratio of the additional wires needed for the coded bus to the number of wires of the original uncoded bus.
- CODEC performance: This is quantified by measuring the complexity, power consumption and speed of the CODEC.
- Bus power: The power saving achieved by the coded bus compared to the uncoded bus.

2.4 Notation and Terminology

Except in Chap. 6, the following notations and terminologies are used unless otherwise specified:

- $V(n)$ denotes an n-bit binary vector. This is sometimes simplified as V. v denotes the value of the vector $V(n)$. v_k denotes the value of the kth bit of V.
- $b_1 b_2 \cdots b_{n-1} b_n$ (or $b_n b_{n-1} \cdots b_2 b_1$) represents an n-bit uncoded input vector (data word). b_n is the most significant bit (MSB) and b_1 is the least significant bit (LSB).
- $d_1 d_2 \cdots d_{m-1} d_m$ (or $d_m d_{m-1} \cdots d_2 d_1$) represents an m-bit coded vector (codeword). d_m is the MSB and d_1 is the LSB.
- A binary string represents the value of a vector. For example: 10101101 represents the value of an 8-bit bus.

- In the case of bus partitioning, a vector V is divided into multiple "groups". Subscripts are used to denote group number. For example. either $(d_1d_2d_3d_4)_1(d_1d_2d_3d_4)_2$ or $d_{1,1}d_{1,2}d_{1,3}d_{1,4}d_{2,1}d_{2,2}d_{2,3}d_{2,4}$ are used to indicate an 8-bit vector partitioned into two groups of 4-bits. $(1011)_1(0101)_2$ means that the patterns in the 1st and 2nd group are 1011 and 0101 respectively.
- If necessary, superscripts are used to denote the cycle number. For example, $(10101101)^k$ is the value of the bus in the kth cycle.
- "$\rightarrow$" denotes a transition. For example, $(10101)^k \rightarrow (01001)^{k+1}$ indicates that the bus pattern transitions from 10101 in the kth cycle to 01001 in the $(k+1)$th cycle.
- "$\Rightarrow$" is an operator that denotes an encoding process. For example, a 3-bit to 4-bit encoding is expressed as $b_1b_2b_3 \Rightarrow d_1d_2d_3d_4$.

Chapter 3
Memoryless Crosstalk Avoidance Codes

Memoryless codes are a class of CACs in which data words and codewords have a 1-to-1 mapping. The encoder uses a fixed codebook and it output is independent of the previously transmitted codewords, hence memoryless.

In this chapter, we present several different types of memoryless codes. These codes offer various degrees of crosstalk reduction that result in a bus speed-up ranging from 100% to over 400%. For each code, we will discuss in detail the code design and quantify its characteristics in terms of speed and area overhead.

3.1 3C-Free CACs

$3C$-free memoryless CACs ensure that $4C$ and $3C$ crosstalk transitions are eliminated from the bus. They are by far the most popular CACs and have been heavily studied in the literature [36, 81, 89, 90, 99, 100]. The bus speed-up they offer is equivalent to passive shielding. In fact, passive shielding can be treated as a $3C$-free memoryless encoding, $b_1 b_2 \cdots b_{n-1} b_n \Rightarrow b_1 \mathbf{0} b_2 \mathbf{0} \cdots \mathbf{0} b_{n-1} \mathbf{0} b_n$. As we will show in this chapter, $3C$-free CACs offer power saving. In addition, Raghunandan et al. pointed out that $3C$-free codes also offers added robustness to process variation [81].

$3C$-free CACs eliminate all *3C and higher* crosstalk patterns in the m-bit bus, and thus satisfy $\max_j(C_{eff,j}) = (1 + 2\lambda)C_L$. From Eq. 2.25, if we assume $\lambda \gg 1$, we know that $\tau_{4C} \approx 2 \cdot \tau_{2C}$. Therefore the maximum speed of a $3C$-free busses is approximately doubled compared to an uncoded bus of the same configuration. Using the actual delay numbers given in Table 2.4, we see that the maximum signal delay on a 20 mm bus with 60× size drivers is reduced from 1561 to 691 ps if $3C$-free encoding is applied. As a result, the maximum speed of such a coded bus is ~1.44 GHz, 125% faster than an uncoded bus.

In the following sections, two types of 3C-free memoryless CACs are discussed: the Forbidden Pattern Free codes and the Forbidden Transition Free codes. We will show that even though these codes are designed based on different principles, their area overhead are very similar. Both types of codes are considered highly efficient.

C. Duan et al., *On and Off-Chip Crosstalk Avoidance in VLSI Design*,
DOI 10.1007/978-1-4419-0947-3_3, © Springer Science+Business Media LLC 2010

3.1.1 Forbidden Pattern Free CAC

The forbidden pattern free codes are the most area overhead efficient 3C-free memory CACs [36]. These codes guarantee that transitions between any two codewords satisfy $2-\delta_{j,j-1}-\delta_{j,j+1} < 3$ for all bits. The detailed design of the code and its performance analysis are given as follows.

3.1.1.1 Code Design

Definition 3.1 *Forbidden patterns* are 3-bit binary patterns $v_{j-1}v_jv_{j+1}$ that satisfy $v_{j-1} = v_{j+1} = \overline{v_j}$.

The only patterns that satisfy Definition 3.1 are *010* and *101*.

Definition 3.2 A vector that contains no forbidden pattern in any three consecutive bits is a *forbidden-pattern-free* (FPF) vector.

For example, 11100110 and 11000110 are FPF vectors, 11<u>010</u>011 and 111<u>101</u>10 are not FPF vectors.

Theorem 3.1 *If all vectors in a set are forbidden pattern free, transitions among these vectors satisfy* $\max_j(C_{eff,j}) = 2 \times C_I$.

Theorem 3.1 states that if forbidden patterns are not allowed on the bus, then $\max(C_{eff}) = 2 \times C_I$, i.e., the bus is 3C-free.

It is possible to prove Theorem 3.1 by an exhaustive examination of all transition patterns. The following gives the mathematical proof.

Proof We prove that both 4C and 3C transitions require forbidden patterns on $d_{j-1}d_jd_{j+1}$. Let the bit patterns be $v_{j-1}^k v_j^k v_{j+1}^k$ and $v_{j-1}^{k+1} v_j^{k+1} v_{j+1}^{k+1}$ in the kth and $(k+1)$th cycles respectively.

- From Eq. 2.24, we know that for $C_{eff,j} = 4 \times C_I$, it is required that $\delta_{j,j-1} = -1$ and $\delta_{j,j+1} = -1$, i.e., v_{j-1} and v_j transit in opposite directions and v_j and v_{j+1} transit in opposite directions. We have the following logical relations:
  - $v_{j-1}^k = \overline{v_j^k}$,
  - $v_{j-1}^k = \overline{v_{j-1}^{k+1}}$,
  - $v_j^k = \overline{v_j^{k+1}}$,
  - $v_{j+1}^k = \overline{v_{j+1}^{k+1}}$ and
  - $v_{j+1}^k = \overline{v_j^k}$

  and these logic relations result in $v_{j-1}^k = \overline{v_j^k} = v_{j+1}^k$ and $v_{j-1}^{k+1} = \overline{v_j^{k+1}} = v_{j+1}^{k+1}$. So in both cycles forbidden patterns appear on $d_{j-1}d_jd_{j+1}$.
- From Eq. 2.24, for $C_{eff,j} = 3 \times C_I$, it is required that $\delta_{j,j-1} = -1$ and $\delta_{j,j+1} = 0$, or $\delta_{j,j-1} = 0$ and $\delta_{j,j+1} = -1$. Without loss of generality, we assume the former case and have

- $v_{j-1}^k = \overline{v_j^k}$,
- $v_{j-1}^k = \overline{v_{j-1}^{k+1}}$,
- $v_j^k = \overline{v_j^{k+1}}$ and
- $v_{j+1}^k = v_{j+1}^{k+1}$

If $v_{j+1}^k = v_j^k$, we will have $v_{j-1}^{k+1} = \overline{v_j^{k+1}} = v_{j+1}^{k+1}$. If $v_{j+1}^k = \overline{v_j^k}$, we have $v_{j-1}^k = \overline{v_j^k} = v_{j+1}^k$. Therefore, a forbidden pattern appears either in the kth cycle or in the $(k+1)$th cycle. □

Combining both cases, we know that by eliminating forbidden patterns on the bus, it is guaranteed that no 4C or 3C transition remain on the bus and $\max(C_{eff}) = 2 \times C_I$.

To generate a complete set of m-bit FPF codewords, a simple-minded approach is to remove all codewords that contains forbidden patterns from the 2^m entries. However, such an exhaustive search approach is impractical when m is large. Instead, the FPF-CAC can be generated using an inductive procedure [36].

Let S_m be the set of m-bit FPF-CAC codewords and let m-bit vector $V_m = d_m d_{m-1} \cdots d_1$ be a codeword. Any codeword $V_m \in S_m$ can be considered as the concatenation of $V_{m-1} = d_{m-1} d_{m-2} \cdots d_1$ with bit d_m, where $V_{m-1} \in S_{m-1}$. The allowed values on d_m is determined by $d_{m-1} d_{m-2}$. For example, if $d_{m-1} d_{m-2} = 00$ or $d_{m-1} d_{m-2} = 11$, then d_m can be either 0 or 1. If $d_{m-1} d_{m-2} = 10$, d_m can only be 0 as $d_m = 1$ will introduce a forbidden pattern, 010, on $d_m d_{m-1} d_{m-2}$.

This idea is utilized in the inductive procedure that generates FPF codewords, given in Algorithm 3.1.

Algorithm 3.1 FPF codeword generation

$S_2 = \{00, 01, 10, 11\}$
for $m \geq 3$ **do**
  $S_m = \emptyset$;
  for $\forall V_{m-1} \in S_{m-1}$ **do**
    if $d_{m-1} d_{m-2} = 00$ or 11 **then**
      add $0 \cdot V_{m-1}$ and $1 \cdot V_{m-1}$ to S_m;
    else if $d_{m-1} d_{m-2} = 01$ or 10 **then**
      add $d_{m-1} \cdot V_{m-1}$ to S_m;
    end if
  end for
end for

"·": concatenation operator

Table 3.1 lists the codewords of the 3, 4 and 5 bit FPF-CACs generated by Algorithm 3.1.

An uncoded n-bit bus has 2^n distinct vectors. For an $m = n + r$ bit bus, there are $T_g(m)$ vectors that are forbidden pattern free. It is necessary that $T_g(m) \geq 2^n$ be

Table 3.1 FPF-CAC codewords for 2, 3, 4 and 5-bit busses

2-bit	3-bits	4-bits	5-bit	
00	000	0000	00000	10000
01	001	0001	00001	10001
10	011	0011	00011	10011
11	100	0110	00110	11000
	110	0111	00111	11001
	111	1000	01100	11100
		1001	01110	11110
		1100	01111	11111
		1110		
		1111		

satisfied so that we can map each input (n-bit) vector uniquely to one distinct output (m-bit) vector that is an FPF vector. To minimize the area overhead, for a given input bus size, n, we want to find the smallest m so that $T_g(m) \geq 2^n$ is satisfied.

The code cardinality, $T_g(m)$, determines the area overhead and therefore is an important parameter in evaluating the code performance. The next section shows the computation of FPF code cardinality.

3.1.1.2 Code Cardinality

From Algorithm 3.1, we can see that for each $m-1$ bit codeword $V_{m-1} \in S_{m-1}$ with last two digits $d_{m-1} = d_{m-2}$, two m-bit codewords can be generated. For V_{m-1} with the last two digits $b_{m-1} \neq b_{m-2}$, only one m-bit codeword can be generated. The total number of FPF-CAC codewords can be computed based on the following equations:

Definition 3.3 For an m-bit vector $d_m d_{m-1} \ldots d_2 d_1$, we define the following quantities:

- $T(m)$ is the total number of distinct m-bit vectors.
- $T_g(m)$ is the total number of FPF vectors.
- $T_b(m)$ is the total number of non FPF vectors.
- $T_{gg}(m)$ is the number of FPF vectors satisfy $d_m = d_{m-1}$.
- $T_{gb}(m)$ is the number of FPF vectors satisfy $d_m \neq d_{m-1}$.

For the base case, a 3-bit bus: $(m = 3)$

$$T(3) = 8,$$

$$T_g(3) = 6,$$

$$T_g(3) = 2,$$

$$T_{gg}(3) = 4,$$

$$T_{gb}(3) = 2$$

For busses with more than 3-bits: ($m > 3$)

$$T_g(m) = 2 \times T_{gg}(m-1) + T_{gb}(m-1) \tag{3.1}$$

$$T_{gg}(m) = T_{gg}(m-1) + T_{gb}(m-1) \tag{3.2}$$

$$T_{gb}(m) = T_{gg}(m-1) \tag{3.3}$$

Based on Algorithm 3.1 and the definitions of T_g, T_{gg} and T_{gb}, we get

$$T_g(m) = T_{gg}(m) + T_{gb}(m) \tag{3.4}$$

$$T_{gg}(m) = T_g(m-1) \tag{3.5}$$

and Eq. 3.1 can be re-written as

$$\begin{aligned} T_g(m) &= 2 \times T_{gg}(m-1) + T_{gb}(m-1) \\ &= (T_{gg}(m-1) + T_{gb}(m-1)) + T_{gg}(m-1) \\ &= T_g(m-1) + T_g(m-2) \end{aligned} \tag{3.6}$$

The relationship shown in Eq. 3.6 is the same as the relationship of elements in the Fibonacci sequence.[1] With the initial conditions $T_g(2) = 4 = 2 \cdot f_3$ and $T_g(3) = 6 = 2 \cdot f_4$, we have

$$T_g(m) = 2 \cdot f_{m+1} \tag{3.7}$$

where f_m is the mth element in the Fibonacci sequence [3].

$T_g(m)$ gives the maximum *cardinality* of the m-bit FPF-CAC code. To encode an n-bit binary bus into FPF-CAC code, the minimum number of bits needed m_{opt} is the smallest integer m that satisfies Eq. 3.8.

$$n \leq \lfloor \log_2 (2 \cdot f_{m+1}) \rfloor \tag{3.8}$$

The mth Fibonacci number can be expressed as

$$f_m = \frac{(-\varphi)^m - (1-\varphi)^m}{\sqrt{5}}, \tag{3.9}$$

where $\varphi = 1.618$ is also known as the *golden ratio*. φ is also the asymptotic ratio of two consecutive elements of the sequence and $\varphi = \lim_{k\to\infty} \frac{f_{k+1}}{f_k}$. For large busses, the lower bound of the area overhead based on Eq. 2.29 is

$$\min(OH(n)) = \frac{m-n}{n} \geq \frac{1}{\log_2 \varphi} - 1 \approx 44\% \tag{3.10}$$

Figure 3.1 plots $\min(OH(n))$ for bus size from 3 to 86 and shows that it asymptotically approaches 44%.

---

[1] A more detailed discussion about Fibonacci sequence will be given in Chap. 4.

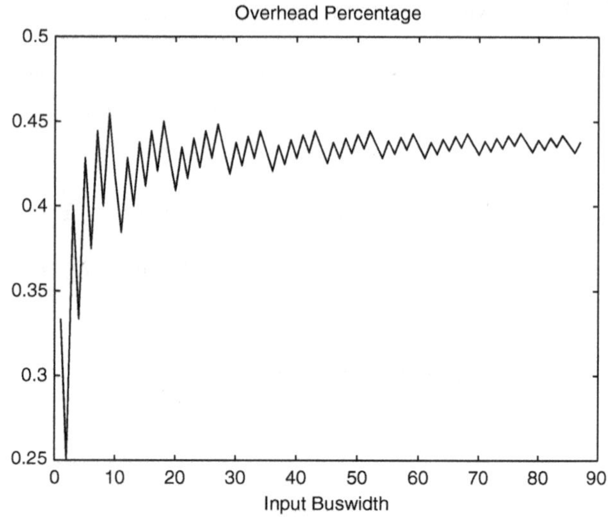

Fig. 3.1 Percentage overhead

If we define the *coding gain* of a CAC, $G(n)$, as the gain of throughput per unit area of the coded bus over the uncoded bus, it can be expressed as

$$G(n) = \frac{R_{enc} \cdot n}{R_{unc} \cdot m} - 1 \tag{3.11}$$

where R_{enc} is the data rate of the encoded bus and R_{unc} is the data rate of the uncoded bus. In the case of 3C-free codes, $R_{enc} \approx 2 \times R_{unc}$. Therefore, the coding gain of the FPF code is

$$G_{FPF} \approx 39\%.$$

Note that the coding gain of a shielding scenario is 0.

3.1.2 Forbidden Transition Free CAC

The forbidden transition free CAC (FTF-CAC) is another efficient $3C$-free memoryless code. It was first proposed by Victor and Keutzer in 2001 [100], and is reviewed in this section since we will use the FTF CAC to derive an efficient CODEC in Sect. 4.5. The basic idea of the FTF code is to prohibit two adjacent bits from transitioning in opposite directions, i.e., the forbidden transitions $01 \rightarrow 10$ or $10 \rightarrow 01$ are not allowed. This guarantees that $\delta_{j,j+1} \geq 0$, therefore $(2 - \delta_{j,j-1} - \delta_{j,j+1}) \leq 2$ and the bus satisfies $\max_j(C_{eff,j}) = 2$. Hence the transition is $3C$-free. An FTF code is a set of codewords such that transitions among these codewords do not produce forbidden transitions on any two adjacent bits.

3.1.2.1 Code Design

A *forbidden transition* is defined as the simultaneous transition (in opposite directions) on two adjacent bits, i.e., 01 → 10 or 10 → 01. In the literature, two adjacent bits are sometimes referred to as a *boundary* and we will use this terminology hereafter. We first observe that to guarantee *forbidden transition freedom* on the boundary $d_j d_{j+1}$ between any two codewords in an FTF-CAC, the 01 and 10 patterns cannot coexist in the same set of codewords. This can be easily confirmed by examining the transitions among codes in $\{00, 01, 11\}$, or $\{00, 10, 11\}$. If we eliminate 01 OR 10 from all the boundaries in the codewords in a set of codewords $\mathbb{R}$, we can guarantee that $\mathbb{R}$ is forbidden transition free. Therefore, once again, the problem of elimination forbidden transitions is transformed into a problem of eliminating specific patterns.

Theorem 3.2 *The largest sets of codewords satisfying the forbidden transition free condition is the set of codewords that can transition to a* class 1 *codeword (defined as a codeword with alternating 0 and 1 bits) without generating forbidden transitions.*

The proof of Theorem 3.2 can be found in [100] and will not be repeated here. For any given size bus, there are two class 1 codewords: "1010..." and "0101...". From Theorem 3.2, we can see that there exist two different sets with the same maximum cardinality. In one set (set A), in all codewords, the 01 pattern is eliminated from $d_{2j+1}d_{2j}$ boundaries and the 10 pattern is eliminated from $d_{2j}d_{2j-1}$ boundaries. In the second set (set B), the 10 pattern is eliminated from $d_{2j+1}d_{2j}$ boundaries and the 01 pattern eliminated from $d_{2j}d_{2j-1}$ boundaries. This is different from the FPF code, which has one unique set with maximum cardinality for each bus width.

Table 3.2 lists all set A FTF codewords for 2, 3, 4 and 5-bit busses. Take the 5-bit bus as an example, we can see that 10 is not present in d_2d_1 and d_4d_3, and 01 is not present in the boundaries d_3d_2 and d_5d_4. If one of these two sets is known, the other set can be produced by simply complementing all the codewords in the first set.

There are multiple methods to produce all the n-bit codewords in the set that satisfy Theorem 3.2.

- Start with a complete set of 2^n vectors and remove codewords that do not satisfy the boundary constraints.

Table 3.2 FTF-CAC codewords for 2, 3, 4 and 5-bit busses

2-bit	3-bits	4-bits	5-bit	
00	000	0000	00000	10100
01	001	0001	00001	10101
11	100	0100	00100	10111
	101	0101	00101	11100
	111	0111	00111	11101
		1100	10000	11111
		1101	10001	
		1111		

- Start with a set consisting of a single class 1 codeword and grow the FTF-CAC codewords by adding compatible codewords to the set.
- Start from a small FTF set (say 2-bit FTF codes) and inductively append bits to the codewords in the set until the codeword length reaches n-bits.

Clearly, the first (pruning) method is impractical when n is large, since a complete set of codewords have 2^n entries and searching through $n-1$ boundaries requires $O(2^{(n-1)n})$ searches.

Both the second and the third methods listed above actually "grow" the FTF codewords instead of "pruning", and therefore require less computation. The method of "growing" codewords by appending bits to codewords in an existing set is similar to the FPF codeword generation algorithm given in Algorithm 3.1. Algorithm 3.2 is the pseudo code for generating the FTF codewords.

Algorithm 3.2 FTF codeword generation

```
S_2 = {00, 01, 11}
for m > 2 do
  if m is odd then
    for ∀V_{m-1} ∈ S_{m-1} do
      add 1 · V_{m-1} to S_m;
      if d_{m-1} = 0 then
        add 0 · V_{m-1} to S_m;
      end if
    end for
  else
    for ∀V_{m-1} ∈ S_{m-1} do
      add 0 · V_{m-1} to S_m;
      if d_{m-1} = 1 then
        add 1 · V_{m-1} to S_m;
      end if
    end for
  end if
end for
```

3.1.2.2 Code Cardinality

The inductive codeword generation method given in Algorithm 3.2 can be used to derive the cardinality of the FTF codes. We first define

Definition 3.4

$$
\begin{aligned}
T_t(m) &: \text{ number of } m\text{-bit FTF vector,} \\
T_{t1}(m) &: \text{ number of } m\text{-bit FTF vector with the MSB being 1,} \\
T_{t0}(m) &: \text{ number of } m\text{-bit FTF vector with the MSB being 0.}
\end{aligned}
$$

The following relationship can be derived from Algorithm 3.2:

$$T_t(m) = T_{t0}(m) + T_{t1}(m) \tag{3.12}$$

$$T_t(2m) = T_{t1}(2m-1) + T_t(2m-1) \tag{3.13}$$

$$T_t(2m-1) = T_{t0}(2m-2) + T_t(2m-2) \tag{3.14}$$

and with some simple manipulation of Eqs. 3.12, 3.13 and 3.14, we get

$$T_t(2m) = T_t(2m-1) + T_t(2m-2), \tag{3.15}$$

$$T_t(2m+1) = T_t(2m) + T_t(2m-1). \tag{3.16}$$

and they can be combined into a single recursive equation

$$T_t(m) = T_t(m-1) + T_t(m-2) \tag{3.17}$$

Once again, Eq. 3.17 resembles the recursive identity of the Fibonacci sequence. Given the initial condition of $T_t(2) = 3 = f_4$ and $T_t(3) = 5 = f_5$, we get

$$T_t(m) = f_{m+2} \tag{3.18}$$

Compare Eq. 3.18 with the maximum cardinality of FPF codes given in Eq. 3.7, we find that the FTF codes have slightly lower cardinality because $2f_{m+1} > f_{m+2} = f_{m+1} + f_m$ for all $m > 0$. However, the asymptotic overhead percentage of the FTF code is still ~44%, the same as the FPF code cardinality given in Eq. 3.10. Therefore for large busses, the coding gain for the FTF-CAC is the same as the coding gain for FPF codes

$$G_{FTF} \approx 39\%.$$

3.1.3 *Circuit Implementation and Simulation Results*

To verify the effectiveness of memoryless crosstalk avoidance coding, we implemented an FPF-CAC bus in SPICE using 0.1 µm BSIM4 model cards [2]. The bus models used were the same as in Sect. 2.2. The coded busses in the simulation were 5-bits wide. A 4-to-5-bit encoder and a 5-to-4-bit decoder logic were manually implemented using an arbitrary mapping of data words to codewords as given in Sect. 4.2. The driver size varied from 30 to 120 times the minimum gate size. Minimum sized inverters are used at the receiver, with a 5 *pF* load (to simulate the gate capacitance of the logic gates driven by the receiver).

Our simulations show that the maximum delay of both the encoder and the decoder was ~25 ps. This added ~50 ps to the overall bus delay (if pipelining is not used). Note that in the case of a pipelined bus, the encoder/decoder delay can be hidden. A random sequence is fed into the input of the bus without the CAC encoders as well as the bus with the FPF-CAC encoder. In the example shown in Fig. 3.2, the

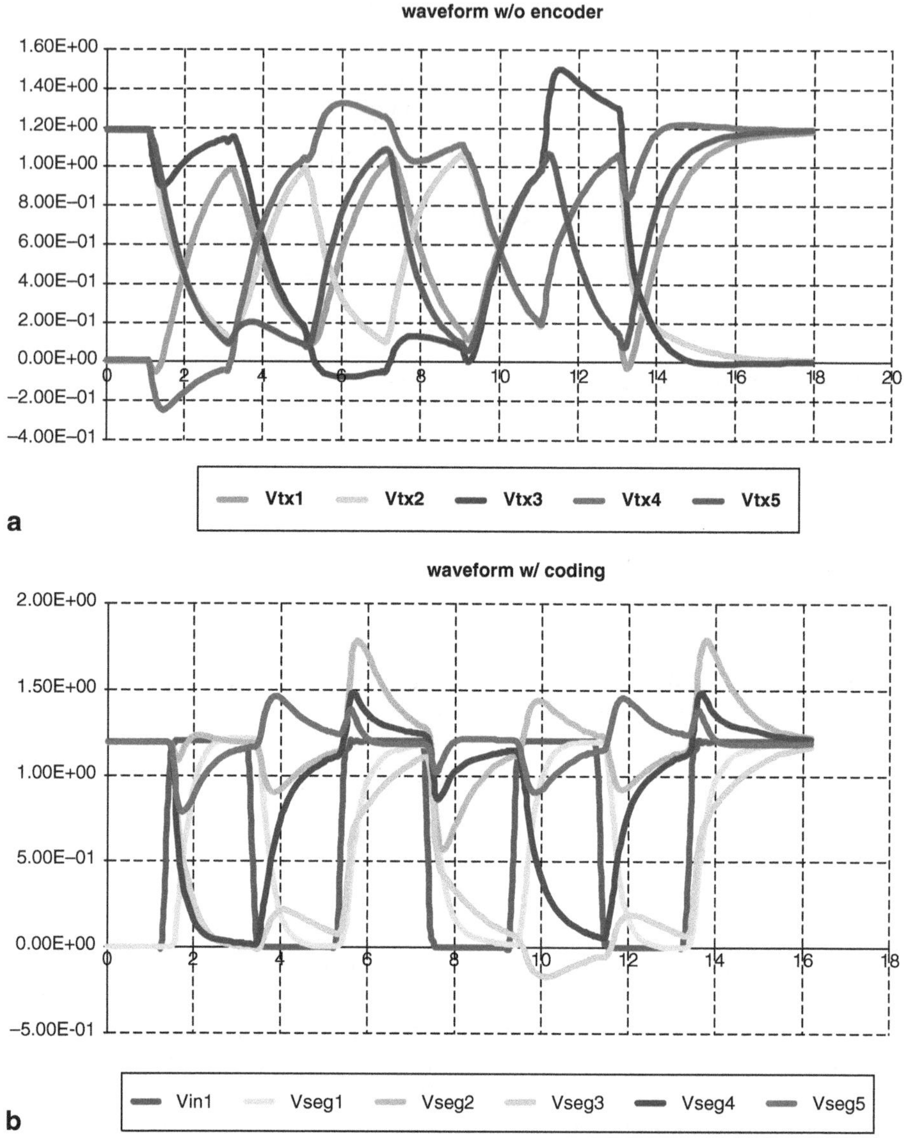

Fig. 3.2 Bus signal waveforms. Bus is 20 mm long and the driver size is 60× the minimum size, implemented in a 0.1 μm process **a** Signal waveform on the bus without coding **b** Signal waveform on the bus with coding

bus lengths are 20 mm and the driver sizes are 60×. The results from the simulation match our theoretical analysis very well. The signal waveforms for an uncoded 4-bit bus in Fig. 3.2a show a worse case delay (from the 50% point at the input of the encoder to the point the bus voltage reaches 50% of V_{dd} at the receiver gates) that is greater than 1000 ps. On the coded bus shown in Fig. 3.2b, the

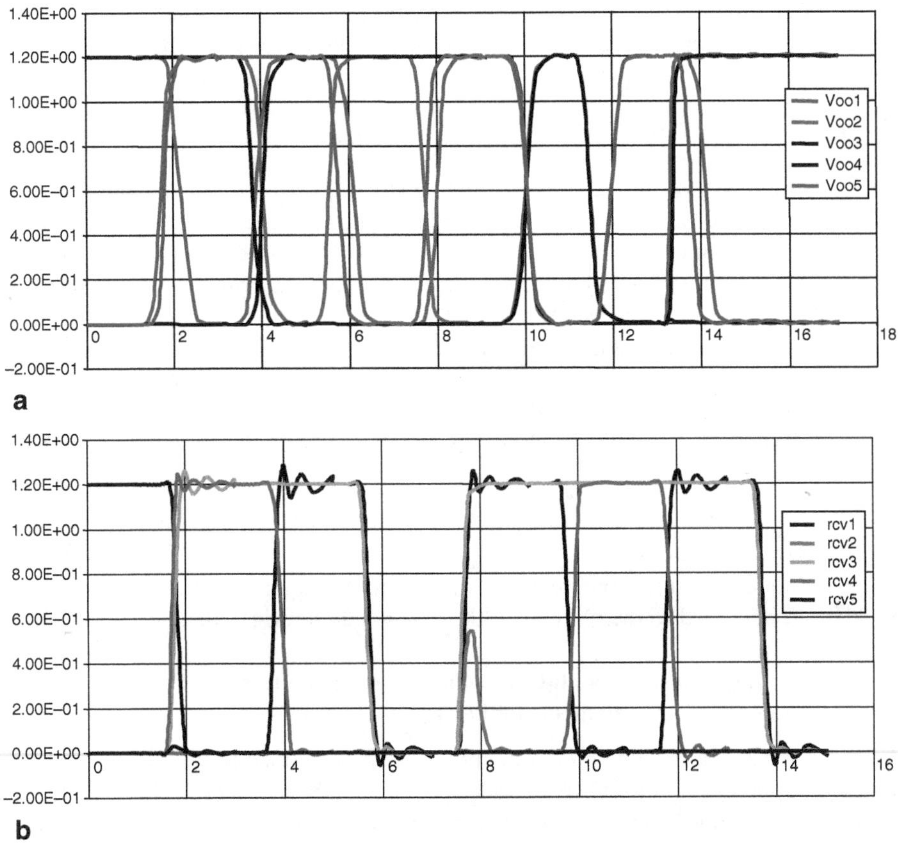

Fig. 3.3 Signal waveforms at the receiver output. Bus is 20 mm long and the driver size is 60× the minimum size, implemented in 0.1 μm process **a** Receiver output waveform on the bus without coding **b** Receiver output waveform on the bus with coding

delay is less than 500 ps. The delay variations are ∼1000 and ∼500 ps respectively. Figure 3.3 shows the reshaped output signals of the receivers. The edge jitter in the uncoded bus and the coded bus are ∼1000 and ∼500 ps respectively. The results indicate that the maximum data rate on the uncoded bus cannot exceed 1 Gbps while a coded bus can operate faster than 2 Gbps, more than doubling the bus speed.

3.2 2C-Free CACs

2*C*-free codes satisfy $\max_j(C_{eff,j}) = (1+\lambda)C_L$ and therefore a 2C-free bus can be significantly faster than a 3*C*-free bus of the same configuration. Using the values from Table 2.4, a 20 mm 2*C*-free bus with 60× drivers can operate at 4.8 GHz, 7.57×

as fast as an uncoded $4C$ bus, or $3.35\times$ as fast as a $3C$-free bus. The $2C$-free codes, however, require larger area overhead as shown in the following sections.

3.2.1 Code Construction

Let Q_n be an n-bit $2C$-free clique (a set of codewords). To satisfy the condition that transitions between any two codewords in the clique are $2C$-free (i.e., $\max_j(C_{eff,j}) = (1+\lambda)C_L$), we know that $C_{lw,j} \leq 1C_I$ and $C_{rw,j} \leq 1C_I$ based on Eq. 2.28, and also $C_{lw,j}$ and $C_{rw,j}$ cannot be $1C_I$ at the same time. Using these criteria, we can examine the properties of all codewords $V_n \in Q_n$.

Again we first examine the 3-bit slice $d_{j-1}d_jd_{j+1}$. We know that "01" and "10" cannot both exist in $d_{j-1}d_j$ boundary as it would violate $C_{lw,j} \leq 1C_I$. Without loss of generality, let us assume that all the valid patterns on $d_{j-1}d_jd_{j+1}$ are in forms of $00\times$, $11\times$ or $01\times$. The value of d_{j+1}, v_{j+1}, shall be chosen based on the following observations:

- if $v_{j+1} = v_j$, the patterns are 000, 111, 011. This satisfy the $2C$-free requirement. This forms a base clique.
- 010 is not compatible[2] with elements in the base clique, 000 and 111.
- 110 can be added to the clique and the expanded clique {000, 111, 011, 110} is still $2C$-free.
- 001 has a $2C$ transition with 011, which is an element in the base clique. It is compatible with other elements in the clique.

An inductive codeword generation method can be developed based on these observations. Suppose an n-bit $2C$-free clique Q_n is available to us, let the $(n+1)$-bit $2C$ crosstalk free clique be initialized to the empty set. We then add $n+1$-bit codewords based on the compatibility requirement as shown in Algorithm 3.3.

Algorithm 3.3 Construction of Q_{n+1} from Q_n

$Q_3 = \{000, 111, 011, 110\}$;
$T(n) = 4$;
$T_{00}(n) = 1$; $T_{01}(n) = 0$; $T_{10}(n) = 1$; $T_{11}(n) = 2$;
$Q_{type} = 10$;
for next n **do**
  $Q_{n+1} = \emptyset$;
  $Q_{n+1} = Q_n \bigcup V_n \cdot v_n, \forall V_n \in Q_n$;
  $T(n+1) = T(n)$;
  $T_{00}(n+1) = T_{00}(n) + T_{10}(n)$;
  $T_{11}(n+1) = T_{11}(n) + T_{01}(n)$;
  $T_{01}(n+1) = 0$; $T_{10}(n+1) = 0$;
  if $Q_{type} = 10$ **then**

[2] By compatible, we mean that two vectors have a non $2C$-free transition.

```
        Q_{n+1} = Q_n ∪ V_{n,00} · 1, ∀V_{n,00} ∈ Q_n;
        T_01(n+1) = T_00(n); T(n+1) = T(n+1) + T_00(n);
        Q_type = 01;
    else
        Q_{n+1} = Q_n ∪ V_{n,11} · 0, ∀V_{n,11} ∈ Q_n;
        Q_type = 10;
        T_10(n+1) = T_11(n); T(n+1) = T(n+1) + T_11(n);
    end if
end for
```

The pseudo-code is given in Algorithm 3.3. The clique type, Q_{type}, is "01" when 01 is present in the last two bits of the codewords in the set. Conversely, $Q_{type} = 10$ when 01 is present in the boundary. $T(n)$ is the cardinality of Q_n and $T_{xy}(0)$ is the number of vectors ending with xy in Q_n. Note that if Q_{type} is "01", (i.e., $T_{10}(n) = 0$), the new clique, Q_{n+1}, derived based on Algorithm 3.3 is of "10" type and therefore does not contain vectors with $d_n d_{n+1} = 10$.

Using the inductive procedure given in Algorithm 3.3, we can produce cliques of arbitrary bus sizes. The procedure, however, does not necessarily yield a $2C$-free clique with maximal cardinality. As we pointed out earlier, vectors $V_{n,00} \cdot 1$ and $V_{01} \cdot 1$ are not compatible, neither are $V_{n,11} \cdot 0$ and $V_{n,10} \cdot 0$. In Algorithm 3.3 we do not add vectors of $V_{n,00} \cdot 1$ and $V_{n,11} \cdot 0$ in the new clique whenever the incompatibility is detected. However, if $V_{n,01} \cdot 1$ vectors are not in the new clique, we can add $V_{n,00} \cdot 1$ vectors into Q_{n+1} instead. Similarly, we can add $V_{n,11} \cdot 0$ vectors instead $V_{n,10} \cdot 0$ vectors into Q_{n+1}. Note that the number of $V_{n,01} \cdot 1$ vectors is $T_{01}(n)$ and the number of $V_{n,00} \cdot 1$ vectors is $T_{00}(n)$. To maximize the size of the new clique, we should select the set with larger number of vectors. The same principle applies to the sets of $V_{n,10} \cdot 0$ vectors and $V_{n,11} \cdot 0$ vectors. Algorithm 3.4 gives the modified inductive codeword generation procedure.

Algorithm 3.4 Modified inductive procedure to construct Q_{n+1} to maximize the clique size

```
Q_3 = {000, 111, 011, 110};
T(n) = 4;
T_00(n) = 1; T_01(n) = 0; T_10(n) = 1; T_11(n) = 2;
Q_type = 10;
for next n do
    Q_{n+1} = ∅;
    Q_{n+1} = Q_n ∪ V_{n,00} · 0, ∀V_{n,00} ∈ Q_n;
    Q_{n+1} = Q_n ∪ V_{n,11} · 1, ∀V_{n,11} ∈ Q_n;
    T(n+1) = T_00(n) + T_11(n);
    T_00(n+1) = T_00(n);
    T_11(n+1) = T_11(n);
    T_01(n+1) = 0; T_10(n+1) = 0;
```

if $Q_{type} == 00$ **then**
  if $T_{00}(n) > T_{11}(n)$ **then**
    $Q_{n+1} = Q_n \bigcup V_{n,00} \cdot 1, \forall V_{n,00} \in Q_n$;
    $T_{01}(n+1) = T_{00}(n)$; $T(n+1) = T(n+1) + T_{00}(n)$;
    $Q_{type} = 01$;
  else
    $Q_{n+1} = Q_n \bigcup V_{n,11} \cdot 0, \forall V_{n,11} \in Q_n$;
    $T_{10}(n+1) = T_{11}(n)$; $T(n+1) = T(n+1) + T_{11}(n)$;
    $Q_{type} = 10$;
  end if
else if $Q_{type} == 10$ **then**
  if $T_{00}(n) > T_{01}(n)$ **then**
    $Q_{n+1} = Q_n \bigcup V_{n,00} \cdot 1, \forall V_{n,00} \in Q_n$;
    $T_{01}(n+1) = T_{00}(n)$; $T(n+1) = T(n+1) + T_{00}(n)$;
    $Q_{type} = 01$;
  else
    $Q_{n+1} = Q_n \bigcup V_{n,01} \cdot 1, \forall V_{n,01} \in Q_n$;
    $T_{11}(n+1) = T_{01}(n)$; $T(n+1) = T(n+1) + T_{01}(n)$;
    $Q_{type} = 00$;
  end if
  else
  if $T_{11}(n) > T_{10}(n)$ **then**
    $Q_{n+1} = Q_n \bigcup V_{n,11} \cdot 0, \forall V_{n,11} \in Q_n$;
    $Q_{type} = 10$;
    $T_{10}(n+1) = T_{11}(n)$; $T(n+1) = T(n+1) + T_{11}(n)$;
  else
    $Q_{n+1} = Q_n \bigcup V_{n,00} \cdot 1, \forall V_{n,00} \in Q_n$;
    $T_{00}(n+1) = T_{10}(n)$; $T(n+1) = T(n+1) + T_{10}(n)$;
    $Q_{type} = 00$;
  end if
end if
end for

3.2.2 Code Cardinality and Area Overhead

For the procedure given in Algorithm 3.3. We have

$$T_{00}(n+1) = \begin{cases} T_{00}(n) + T_{10}(n), & \texttt{if } T_{01}(n) = 0, \\ T_{00}(n), \texttt{otherwise} \end{cases} \tag{3.19}$$

$$T_{11}(n+1) = \begin{cases} T_{11}(n), & \texttt{if } T_{01}(n) = 0, \\ T_{11}(n) + T_{01}(n), & \texttt{otherwise} \end{cases} \tag{3.20}$$

$$T_{01}(n+1) = \begin{cases} T_{00}(n), \ \texttt{if} \ T_{01}(n) = 0, \\ 0, \ \texttt{otherwise} \end{cases} \tag{3.21}$$

$$T_{10}(n+1) = \begin{cases} 0, \ \texttt{if} \ T_{01}(n) = 0, \\ T_{11}(n), \ \texttt{otherwise} \end{cases} \tag{3.22}$$

Assume $Q_{type}(n)$ is "10", i.e. $T_{01}(n) = 0$, according to Algorithm 3.3, both $Q_{type}(n+1)$ and $Q_{type}(n-1)$ are "01". Equations 3.19, 3.20, 3.21 and 3.22 can be simplified into

$$T_{00}(n+1) = T_{00}(n-1) + T_{11}(n-1) \tag{3.23}$$

$$\begin{aligned} T_{11}(n+1) &= T_{11}(n-1) + T_{01}(n-1) \\ &= T_{11}(n-1) + T_{00}(n-3) \end{aligned} \tag{3.24}$$

$$T_{01}(n+1) = T_{00}(n-1) \tag{3.25}$$

$$T_{10}(n+1) = 0. \tag{3.26}$$

and we also have

$$\begin{aligned} T(n+1) &= T_{00}(n+1) + T_{11}(n+1) + T_{01}(n+1) + T_{10}(n+1) \\ &= T_{00}(n-1) + T_{11}(n-1) + T_{11}(n-1) + T_{01}(n-1) + T_{00}(n-1) \\ &= 2 \cdot T(n-1) - T_{01}(n-1) \\ &= 2 \cdot T(n-1) - T_{00}(n-3) \end{aligned} \tag{3.27}$$

Similar equations can be derived for $Q_{type}(n)$ of "01" and we will not list the details here. Since $Q_2 = \{00, 11, 01\}$, the initial conditions are

$$\begin{aligned} T_{00}(2) &= 1, \\ T_{11}(2) &= 1, \\ T_{01}(2) &= 1, \\ T_{10}(2) &= 0. \end{aligned}$$

The closed-form expression for the code cardinality is not obvious. Instead, the numerical results can be easily obtained from Eqs. 3.23, 3.24 and 3.27. The computed results are shown in Fig. 3.4. This figure describes the relationship between the actual bus size (y-axis) and the effective bus size (x-axis) which is given by $m = \lfloor \log_2(T_n) \rfloor$. This is plotted using a dotted line. The percentage overhead is shown as well (solid line). We note that the asymptotic overhead is about 146%. The corresponding coding gain is

$$G_{2CF} = \frac{R_{1C}}{R_{4C} \times 2.46} - 1 \approx \frac{4}{2.46} - 1 = 62\%$$

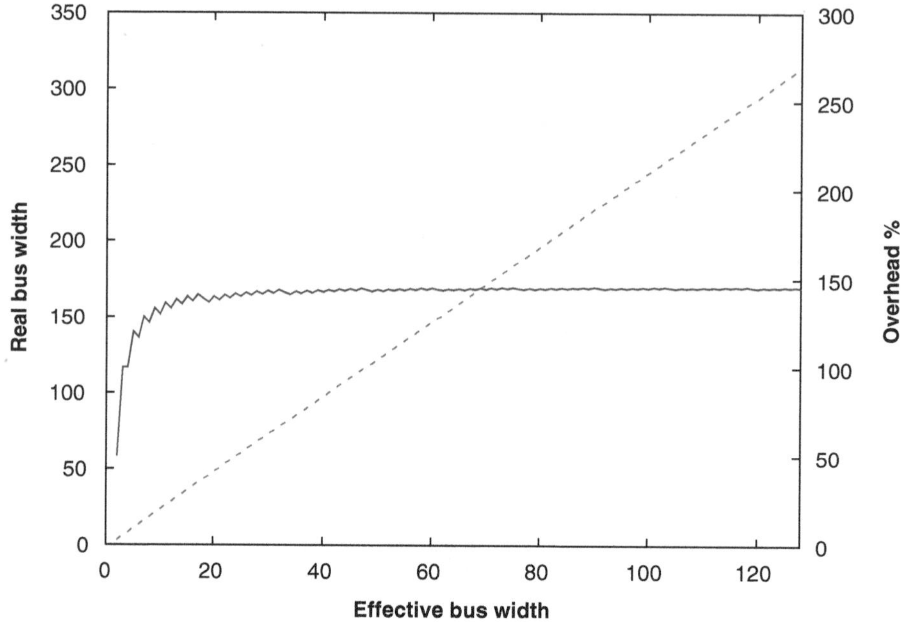

Fig. 3.4 Effective versus actual bus width for $2C$-free CACs

3.2.3 2C Experiments

We implemented our $2C$ crosstalk free CODEC using PLAs based on the codewords generated using Algorithm 3.3. In particular, we implemented a $3 \Rightarrow 6$-bit encoder, and determined its delay to be 179 ps, and its area to be 10 μm^2. For larger buses, we would perform the encoding by concatenating an appropriate number of such encoders. When adjacent bits of adjacent encoded buses have an opposite value, we complement the encoded vector of one of these buses, avoiding $2C$ crosstalk between the encoded buses.

It is worth mentioning that in this case, the CODEC speed can be the limiting factor on system performance, since many entries under the column $1C$ of Table 2.4 report a delay less than the encoder delay. In spite of this, the speedup of the data transfer is quite significant (approximately between $3\times$ and $7\times$).

3.3 1C-Free Busses

A $1C$-free bus requires $\delta_{j,j\pm1} = 1$. This indicates that two neighboring wires of the jth wire must transition in the same direction. The only coding scheme that satisfies this requirement is to duplicate each input wire 3 times, i.e., $d_{j-1}d_jd_{j+1} = b_kb_kb_k$. This scheme incurs a fixed area overhead of 200%. The coding in this case is trivial. We compare the bus performance in different physical configurations. The variables

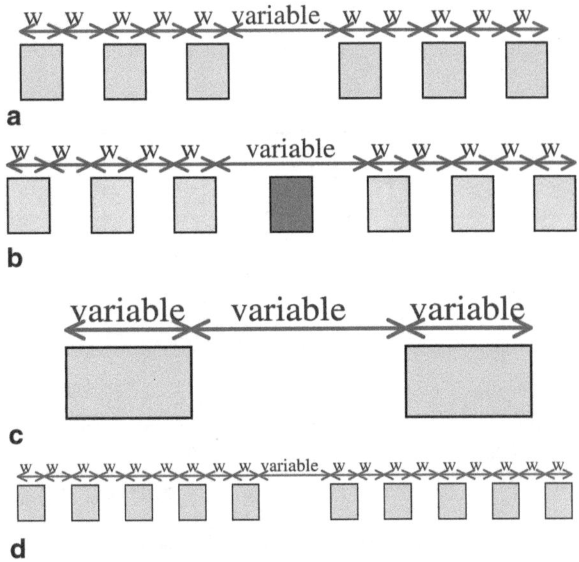

Fig. 3.5 $1C$ free bus configurations **a** 3-wire group, fixed spacing within group, variable spacing between groups **b** 3-wire group with shielding between groups, fixed spacing within group, variable spacing between groups **c** no shielding wires, variable wire sizes and spacing **d** 5-wire group, fixed spacing within group, variable spacing between groups

in different configurations include the number of duplicated wires, spacing between wires, wire line width and whether or not shielding wires are used.

3.3.1 Bus Configurations

We compare 4 different configurations to eliminate $1C$ crosstalk. These configurations are illustrated in Fig. 3.5.

- Configuration (a) consists of groups of 3 minimum width wires for each bus signal, separated by minimum spacing. Groups of these wires are separated by a variable distance.
- Configuration (b) is similar to (a) with the exception that groups of bus signal wires are separated by a GND signal.
- Configuration (c) consists of a single wire (of varying width) for each bus signal, with adjacent bus signal wires separated by a variable spacing.
- Configuration (d) is similar to (a) with the difference that each bus signal consists of a group of 5 minimum width wires separated by minimum spacing. Groups of these wires are separated by a variable distance.

In configurations (a), (b) and (d), the signal of the central wire of each group is used by the receiving circuitry.

3.3.2 Experimental Results

Figure 3.6 compares the speed and overhead performance for all the configurations shown in Fig. 3.5. The delays in this figure correspond to the case where neighboring

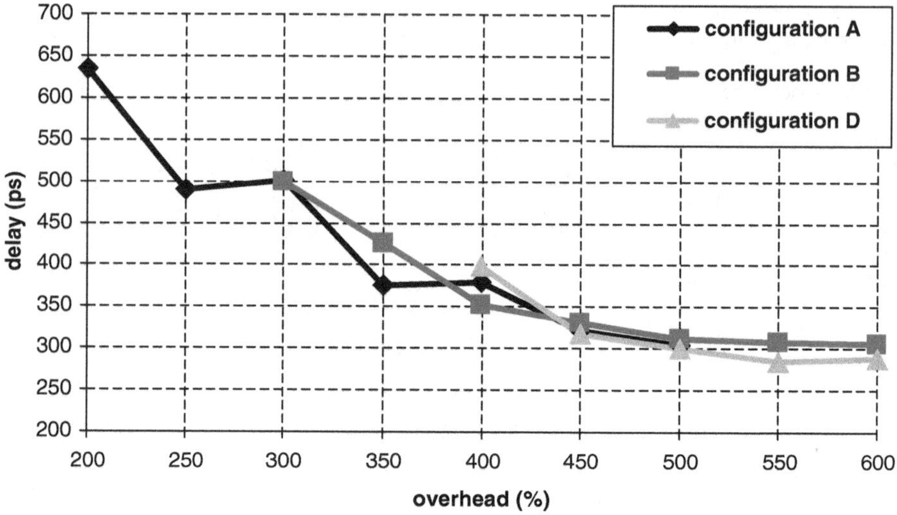

Fig. 3.6 Delay versus area tradeoff for 1C schemes

groups of 3 wires (which represent the same bus signal) switch in opposite directions (resulting in the maximum delay). The length of wires was 20 mm, and the size of drivers was 30× minimum.

Configuration (c) resulted in significantly larger delay penalties that the other configurations, because of the absence of any shielding. Such a configuration violates the $1C$-free condition and therefore has the worst performance. Among the other configurations, we note that the minimum delay is *greater* than the delay of $0C$ signals given in Table 2.4. This is because it is exceedingly hard to ensure that all 3 (or 5) wires (representing the same bus signal) switch at exactly the same time as required for a $0C$ transition. The possible skew induced on the outer shielding wire is sufficiently large to cause the transition of the center wire to effectively have close to $2C$ crosstalk. This was verified by simulations where we intentionally skewed the outer and central signals. These experiments indicate that a $1C$-free CAC is not practical.

The "1C-free" bus describes the crosstalk properties of the busses. Readers may notice that some of these scheme discussed above is very similar to active shielding [60]. In fact, the active shielding is the implementation of the "1C free" coded bus. These two terms can be used inter-changeably.

3.4 Summary

We discussed several memoryless CACs in this chapter. These codes offer speed-up factors from 100 to 400%. Based on our analysis, we can see that in general, there is a trade-off between speed-up factor and area overhead, the higher the speed-up factor, the higher the area overhead incurs. However, we showed that carefully designed codes can shift the balance point of this trade-off and provides a significant coding gain. We proved that both the FPF and the FTF encoding schemes can be used for a $3C$-free bus and they have slightly different but asymptotically identical overhead performance. We gave the inductive algorithms for the generation of FPF and FTF codes. Two algorithms were also given for inductively generating $2C$-free codewords. The percentage overhead for $2C$-free CACs was computed numerically. Finally, we compared the performance of different $1C$-free bus configurations. We showed simulation results that for a given area overhead, all configurations yield virtually the same amount of speed-up, which is lower than what is expected in theory.

Chapter 4
CODEC Designs for Memoryless Crosstalk Avoidance Codes

The 1C-free bus described in Sect. 3.3 does not require CODECs, or we can say that these CODEC designs are trivial (repeating the input bit by N times). On the other hand, 3C-free (Sect. 3.1) and 2C-free (Sect. 3.2) codes need CODECs. For these codes to be used in practice, efficient CODEC designs are necessary. In the case of crosstalk avoidance codes, the complexity and speed of the CODEC are both critical for overall bus performance. Also, the power consumption of the CODECs should be factored in when the overall power consumption is evaluated. In this chapter, we present several efficient CODEC design techniques for the memoryless CACs. The advantages of these CODECs are their low complexity, high speed as well as minimum area overhead.

The first two techniques described in this chapter are the "group complement" [36] and "bit overlapping" [89] approaches. Both are based on the idea of bus partitioning. Even though the examples given are CODECs for 3C-free codes, the general idea can be easily extended to 2C-free codes. Two other CODECs in this chapter are for FPF codes and FTF codes respectively and both are based on *Fibonacci Numeral System*-based mapping scheme.

4.1 Bus Partitioning Based CODEC Design Techniques

A crosstalk avoidance CODEC, when implemented as a flat design without hierarchy, is expected to exhibit an exponential growth in area with the size of the bus. This was confirmed by Sridhara et al. in [89]. Figure 4.1 shows the gate count of their implementation of a look-up table based FTF encoder for input bus sizes from 3-bits to 12-bits. We can see that at 12-bits, the total gate count exceeds 20 K. Clearly, the added area and energy consumption of the encoder/decoder will quickly cancel out any saving that the coding offers. We can also expect the CODEC speed to degrade rapidly as the bus grows, to an extent where it could dominate the crosstalk-induced bus delay and hence become the speed bottleneck for the bus.

The basic idea of bus partitioning is to divide a large bus into small *groups* (*lanes*) and encode each group independently. By limiting the size of each group (typically 3–6 bits), we can keep the CODEC of each group small and hence it can operate at

C. Duan et al., *On and Off-Chip Crosstalk Avoidance in VLSI Design*,
DOI 10.1007/978-1-4419-0947-3_4, © Springer Science+Business Media LLC 2010

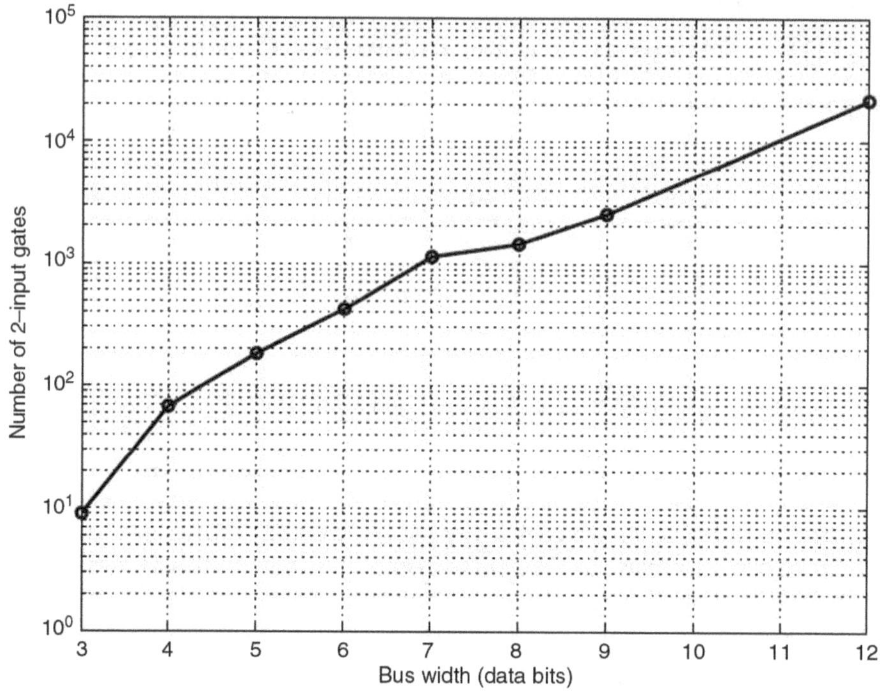

Fig. 4.1 Gate count of flat look-up table based FTF encoder designs [89]

high speed. The difficulty arises when merging all the encoded groups into a single bus, since undesired crosstalk transitions often occur between adjacent groups. The following are two techniques designed to address this issue.

4.2 Group Complement

The discussion of code design in Sect. 3.1.1.1 does not specify the mapping between the input data words and the output codewords. Table 4.1 shows a particular mapping for the 4 $\Rightarrow$ 5-bits FPF code. This mapping is chosen arbitrarily and we cannot easily verify if it yields the most efficient implementation. It is quite possible that more efficient mappings exist, given that there are a total of 16! $\sim 2 \times 10^{13}$ possible mappings in all. This prohibits us from performing an exhaustive search to find the optimal CODEC.

To reduce the problem size, a group complement FPF encoder can be implemented as shown in the example of a 16-bit input in Fig. 4.2. The input is divided into four 4-bit groups and the data of each group is encoded using a 4 $\Rightarrow$ 5-bits encoder. Based on the previous discussion, all $2^4 = 16$ vectors of the 4-bit bus can be mapped to 16 5-bit $3C$-free FPF vectors. There are no forbidden patterns within any of the 5-bit vectors output by the encoder. To achieve the optimum, we then still need to find

Table 4.1 A 4 ⇒ 5-bit FPF encoder input-output mapping

Input b_1	b_2	b_3	b_4	Output d_1	d_2	d_3	d_4	d_5
0	0	0	0	0	0	0	0	0
0	0	0	1	0	0	0	0	1
0	0	1	0	0	0	1	1	0
0	0	1	1	0	0	0	1	1
0	1	0	0	0	1	1	0	0
0	1	0	1	0	0	1	1	1
0	1	1	0	0	1	1	1	0
0	1	1	1	0	1	1	1	1
1	0	0	0	1	1	1	1	1
1	0	0	1	1	1	1	1	0
1	0	1	0	1	1	0	0	1
1	0	1	1	1	1	1	0	0
1	1	0	0	1	0	0	1	1
1	1	0	1	1	1	0	0	0
1	1	1	0	1	0	0	0	1
1	1	1	1	1	0	0	0	0

the mapping that minimizes the CODEC size by searching exhaustively through all possible mapping.

Fortunately for such a small bus size, the encoder and decoder sizes are very small even without an optimal mapping. We can either search for a small subset of all possible mapping, or randomly choose a mapping and rely on synthesis tools to generate the CODEC logic for us. The encoder logic corresponding to a randomly selected mapping given in Table 4.1 is:

$$d_1 = b_1,$$

$$d_2 = (b_2b_3 + b_2\overline{b_4}) \oplus b_1,$$

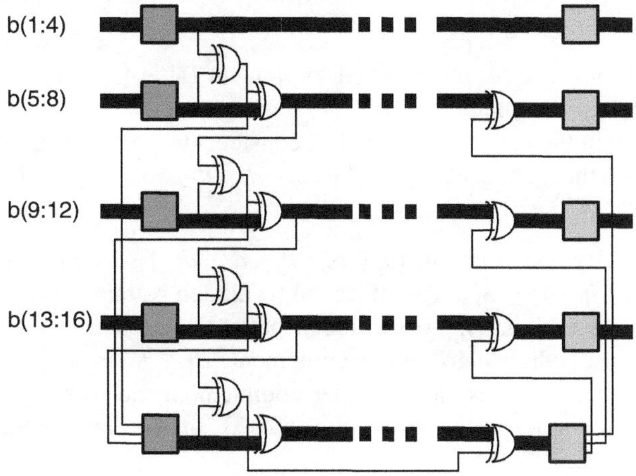

Fig. 4.2 A 16-bit group complement FPF CODEC structure

$$
\begin{aligned}
d_3 &= (b_2 + b_3\overline{b_4}) \oplus b_1, \\
d_4 &= (b_3 + b_2 b_4) \oplus b_1, \\
d_5 &= b_4 \oplus b_1
\end{aligned}
\tag{4.1}
$$

and the decoder logic is

$$
\begin{aligned}
b_1 &= d_1 \\
b_2 &= ((d_2\overline{d_5} + d_4 d_5) d_3) \oplus d_1 \\
b_3 &= ((d_3\overline{d_5} + d_2 d_5) d_4) \oplus d_1 \\
b_4 &= d_5 \oplus d_1
\end{aligned}
\tag{4.2}
$$

When combining groups of the 16-bit bus, however, forbidden pattern can occur **across the group boundaries**. For instance, consider two encoded adjacent groups with data {111**10**}{**11**100}. Now there is a forbidden pattern between these groups (shown underlined). To prevent this from occurring, we invert the outputs of the second encoder if $v_{j,5} \neq v_{j+1,1}$ where j is the group index. The final output for our example will therefore be $\{11110\}\{00011\}$. An **group complement bit** is also transmitted, to enable the decoder to correctly decode the transmitted vector. The entire set of group complement bits are transmitted in a separate group, as shown in Fig. 4.2.

In the above example, we need a total of 26 bits to be transmitted (for a 16-bit bus). This give an overhead of 62.5%, higher than the lower bound of 44%.

4.2.1 Proof of Correctness

Definition 4.1 S_{v5} is defined as the set of all possible 5-bit FPF vectors.

Theorem 4.1 *A bus partitioned using the group complement approach is* 3*C-free if each of its group is* 3*C-free.*

Proof The column marked "output" in Table 4.1 is the set S_{v5}. Note that S_{v5} is closed under complement.

In the encoder of Fig. 4.2, consider two adjacent groups of encoded bus signals. Let these encoded signals be $\{d_{j,1}, d_{j,2}, d_{j,3}, d_{j,4}, d_{j,5}\}$ and $\{d_{j+1,1}, d_{j+1,2}, d_{j+1,3}, d_{j+1,4}, d_{j+1,5}\}$.

- By construction, $\{d_{j,1}, d_{j,2}, d_{j,3}, d_{j,4}, d_{j,5}\} \in S_{v5}$ and $\{d_{j+1,1}, d_{j+1,2}, d_{j+1,3}, d_{j+1,4}, d_{j+1,5}\} \in S_{v5}$. Therefore, no forbidden pattern occurs in either of these groups.
- Additionally, no forbidden pattern occurs across group j and $j+1$ since the encoding algorithm complements $d_{j+1,i} (1 \leq i \leq 5)$ whenever $d_{j,5} \neq d_{j+1,1}$. Since S_{v5} is closed under complement, no forbidden patterns result within the complemented group either. Also, since complement ensures that $d_{j,5} = d_{j+1,1}$

in the transmitted data, no forbidden transitions can occur across groups j and $j+1$.

We shall note that the group complement approach does not apply to FTF codes because these codes are not closed under complement. □

4.3 Bit Overlapping

Bit overlapping was first proposed by Sridhara et al. in [89]. In this method, an input bus is divided into groups. However, between two adjacent groups, an overlap bit is used, i.e., a bit appearing as the last bit of one group will also be the first bit of the following group. For example, two groups are $b_{k-G+1} \cdots b_{k-1}b_k$ and $b_k b_{k+1} \cdots b_{k+G-1}$, where G is the group size. A mapping is chosen so that the forbidden pattern across the boundaries are avoided.

The bit overlapping technique can be applied to both FPF and FTF codes. Figure 4.3 illustrate the design of a FPF bit-overlapping encoder. In both cases, the resulting CODECs have higher are overhead than the theoretical lower bound. Interested readers can refer to [89] for more design details.

4.4 FPF-CAC CODEC Design

Both bus partitioning based designs yield small and fast CODECs. Unfortunately, they are not efficient because of the additional wires required to handle crosstalk across adjacent groups. In this section, we propose two CODEC designs that allow us to encode data to the FPF-CAC *without* partitioning the bus. The mapping schemes allow us to systematically construct the FPF-CAC or FTF-CAC CODECs for busses of arbitrary size. By "systematically", we mean that the CODEC for a larger bus is

Data	Code
0000	00000
0001	00001
0010	00110
0011	00011
0100	01100
0101	00111
0110	01110
0111	01111
1000	10000
1001	10001
1010	11000
1011	10011
1100	11100
1101	11001
1110	11110
1111	11111

a

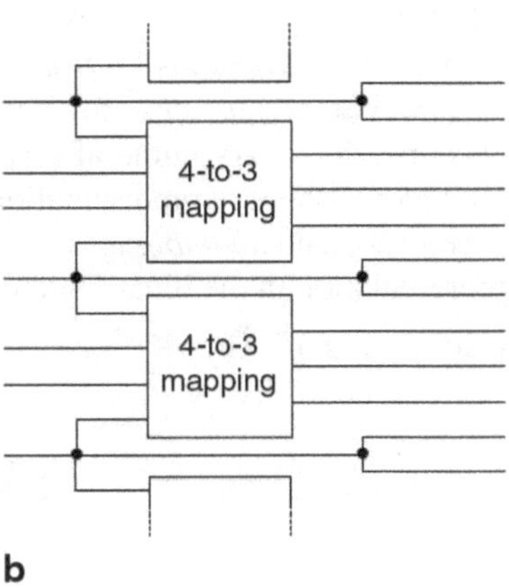

b

Fig. 4.3 FPF bit-overlapping encoder

obtained by extending the CODEC of a smaller bus. The gate counts of the proposed CODEC implementations grow approximately quadratically with respect to the bus size, instead of exponentially for previous approaches [89].

We have shown in Sect. 3.1.1.1 that 3C and 4C crosstalk classes can be avoided if the bus is encoded using the FPF code. We provided the recursive procedure for generating the FPF codewords and provided a procedure to compute the total number of codewords and the lower bound for the bus area overhead. However, we did not discuss the mapping scheme between the input data words and the output codewords, neither did we show how a CODEC for the FPF-CAC can be constructed.

Conceptually, the mapping between data words and codewords is flexible, provided it can be reversed by the decoder. When the size of the code book S is a power of 2, $S!$ CODEC implementations are possible. In the case when the size of the code book is not a power of two, a 1-to-1 mapping is not required. A 1-to-many mapping for certain data words may improve the CODEC performance further. Finding the optimal CODEC, however, turns out to be even more difficult given the large number of permutations, the large flexibility in choosing the codewords of the CODEC, and the non-linear relationship between the CODEC characteristics (area, delay and power) and the input-output mapping.

In the remainder of this section, we describe how a Fibonacci numeral system based mapping yields a systematic method to design an efficient and predictable CODEC.

4.4.1 *Fibonacci-Based Binary Numeral System*

The Fibonacci binary numeral system (FNS) was first mentioned in the context of CAC designs by Mutyam in [75]. The author proposed an inductive codeword generation algorithm for the forbidden transition free code. The algorithm is similar to those proposed in [36, 100]. However, [75] failed to address the mapping scheme and CODEC design. We next describe our FNS-based mapping, and the resulting CODEC designs.

A numeral system is *"a framework where numbers are represented by numerals in a consistent manner"* [5]. The most commonly used numeral system in digital design is the binary numeral system, which uses powers of two as the *basis*. For a number v, its binary representation is defined in Eq. 4.3. The binary numeral system is *complete* and *unambiguous*, which means that each number has one and only one representation in the binary numeral system.

Definition 4.2

$$v = \sum_{k=1}^{n} b_k \cdot 2^{k-1} \qquad b_k \in \{0,1\} \tag{4.3}$$

$$= \sum_{k=1}^{m} d_k \cdot f_k \qquad d_k \in \{0,1\} \tag{4.4}$$

The Fibonacci-based numeral system $N(F_m, \{0,1\})$ is the numeral system that uses Fibonacci sequence as the basis. The definition of the Fibonacci sequence [3] is given in Eq. 4.5. A number v is represented as the summation of some Fibonacci numbers, and no Fibonacci number is in the summation more that once, as indicated in Eq. 4.4.

Definition 4.3

$$f_m = \begin{cases} 0 & \text{if } m = 0, \\ 1 & \text{if } m = 1, \\ f_{m-1} + f_{m-2} & \text{if } m \geq 2. \end{cases} \tag{4.5}$$

Similar to the binary numeral system, the Fibonacci-based numeral system is complete, and therefore any number v can be represented in this system. However, the Fibonacci-based numeral system is *ambiguous*. As an example, there are *six* 7-digit vectors in the Fibonacci numeral system for the decimal number 19: {0111101, 0111110, 1001101, 1001110, 1010001, 1010010}. For clarity, we refer to a vector in the binary numeral system as a *binary vector* or *binary code*; a vector in the Fibonacci numeral system is referred to as a *Fibonacci vector* or *Fibonacci code*. All the Fibonacci vectors that represent the same value are defined as *equivalent vectors*.

Another very important identity of the Fibonacci sequence that is used in the following discussions is

$$f_m = \sum_{k=0}^{m-2} f_k + 1 \tag{4.6}$$

The n-bit binary vector can represent numbers in the range of $[0, 2^n - 1]$, and therefore a total of 2^n values can be represented by n-bit binary vectors. From Eq. 4.6, we know that the range of an m-bit Fibonacci vector is $[0, f_{m+2} - 1]$, where the minimum value 0 corresponds to all the bits d_k being 0, and the maximum value corresponds to all d_k being 1. Hence a total of f_{m+2} distinct values can be represented by m-bit Fibonacci vectors.

4.4.2 Near-Optimal CODEC

We first propose a coding scheme that converts the input data to a forbidden pattern free Fibonacci vector. The code is near-optimal since the required overhead is no more than 1 additional bit, compared to the theoretical lower bound given in Eq. 3.10.

The coding algorithm is developed based on a result that states that any number v can be represented in FNS, in an FPF manner. In order to prove this result, we first derive the following corollaries:

Corollary 4.1 *The following two m-bit Fibonacci vectors are equivalent:* $d_m d_{m-1} \cdots d_3 01$ *and* $d_m d_{m-1} \cdots d_3 10$. *In other words,* $d_m d_{m-1} \cdots d_3 01 \equiv d_m d_{m-1} \cdots d_3 10$.

Proof Since $f_2 = f_1 = 1$, it is obvious that the last two digits are interchangeable. □

Corollary 4.2 *For an m-bit Fibonacci vector* $d_m d_{m-1} \cdots d_2 d_1$*, if three consecutive bits* $d_k d_{k-1} d_{k-2}$ *have a value* 100*, replacing them with* 011 *produces an equivalent vector.*

Proof This can be proven based on the definition of Fibonacci sequence, given in Eq. 4.5. □

Corollary 4.3 *For an m-bit Fibonacci vector* $d_m d_{m-1} \cdots d_2 d_1$*, if a number of consecutive bits (slice)* $d_k d_{k-1} \cdots d_{k-n}$ *in the vector have a pattern of* $0101 \cdots 0100$ *(alternating 0 and 1 except for the last two bits), then replacing the slice with the pattern* $0011 \cdots 1111$ *(all bits are 1 except the first two bits) produces an equivalent vector.*

Proof From the right to left, we can recursively replace 100 with 011 (Corollary 4.2) until $d_{k-1} d_{k-2} d_{k-3}$ has been replaced and the slice is modified to $0011 \cdots 111$. □

Corollary 4.4 *For an m-bit Fibonacci vector* $d_m d_{m-1} \cdots d_1$*, if a slice in the vector* $d_k d_{k-1} \cdots d_{k-n}$ *has a pattern of* $1010 \cdots 1011$ *(alternating 1 and 0 except the last two bits), replacing the slice with the pattern* $1100 \cdots 0000$ *(first two bits are 1s and the other bits are 0s) produces an equivalent vector.*

Proof Similar to the proof of Corollary 4.3. From the right to left, we recursively replace 011 with 100 (Corollary 4.2) until $d_{k-1} d_{k-2} d_{k-3}$ has been replaced, and the slice is transformed to $1100 \cdots 000$. □

With all the above corollaries, we now state and prove the following theorem.

Theorem 4.2 $\forall v \in [0, f_{m+2} - 1]$, $\exists d_m d_{m-1} \cdots d_2 d_1$*, S.T.* $v = \sum_{k=1}^{m} d_k \cdot f_k$ *and* $d_m d_{m-1} \cdots d_2 d_1$ *is FPF.*

Theorem 4.2 states that for any number $v \in [0, f_{m+2} - 1]$, there exists at least one m-bit Fibonacci vector $d_m d_{m-1} \cdots d_2 d_1$ such that this vector represents v and is *forbidden pattern free*.

Proof

- $\forall v \in [0, f_{m+2} - 1]$, $\exists d_m d_{m-1} \cdots d_2 d_1$, S.T. $v = \sum_{k=1}^{m} d_k \cdot f_k$.
  In other words, for any number $v \in [0, f_{m+2} - 1]$, there exists at least one m-bit Fibonacci vector $d_m d_{m-1} \cdots d_1$. This follows from the completeness of the Fibonacci number system.
- If the vector $d_m d_{m-1} \cdots d_1$ is *not* an FPF vector, we can produce at least one equivalent vector that is FPF by performing some or all of the following procedures:
  - If the vector ends with a forbidden pattern (101 or 010), there exists an equivalent vector that ends with 110 or 001 (Corollary 4.1).
  - If any slice in the vector contains a forbidden pattern, they can be replaced with a slice that is forbidden pattern free using Corollary 4.3 and 4.4. □

By proving Theorem 4.2, we show the existence of an FPF Fibonacci vector for any number v in the range $[0, f_{m+2} - 1]$. The coding algorithm that encodes a given number v to an FPF Fibonacci vector is given in Algorithm 4.1.

Algorithm 4.1 Near-Optimal FPF-CAC encoding algorithm: FPF-CAC(v)

```
\\ MSB stage:
if v ≥ f_{m+1} then
    d_m = 1;
    r_m = v − f_m;
else
    d_m = 0;
    r_m = v;
end if
\\ other stages:
for k = m-1 to 2 do
    if r_{k+1} ≥ f_{k+1} then
        d_k = 1;
    else if r_{k+1} < f_k then
        d_k = 0;
    else
        d_k = d_{k+1};
    end if
    r_k = r_{k+1} − f_k · d_k;
end for
\\LSB
d_1 = r_2;
return (d_m d_{m−1} ··· d_1);
```

Algorithm 4.1 shows that an m-bit FPF vector is generated in m stages. Each stage outputs one bit of the output vector (d_k) and the remainder (r_k) that is the input to the following stage. In the kth stage, the input r_{k+1} is compared to two Fibonacci numbers f_{k+1} and f_k. If $r_{k+1} \geq f_{k+1}$, d_k is coded as 1; If $r_{k+1} < f_k$, d_k is coded as 0; If the value r_{k+1} is in between, d_k is coded to the same value as d_{k+1}. The remainder is computed as $r_{k+1} - d_k \cdot f_k$. We shall refer the ranges $[f_{k+1}, f_{k+2})$, (f_k, f_{k+1}) and $[0, f_k)$ as the ***force-1 zone***, ***gray zone*** and ***force-0 zone*** of the kth stage respectively. The most significant bit (MSB) stage is slightly different from other stages since no bit proceeds it. It encodes d_m by comparing the input v with only one Fibonacci number, f_{m+1}.

The decoder is a straightforward implementation of Eq. 4.4 which converts the Fibonacci vector back to the binary vector. Figure 4.4 shows the encoder and decoder structures based on Algorithm 4.1.

The correctness of Algorithm 4.1 can be proven by showing that if after the kth stage, the partially generated output vector $d_m \cdots d_{k+1} d_k$ is FPF, then adding the output of the $(k-1)$th stage, d_{k-1} will not introduce a forbidden pattern.

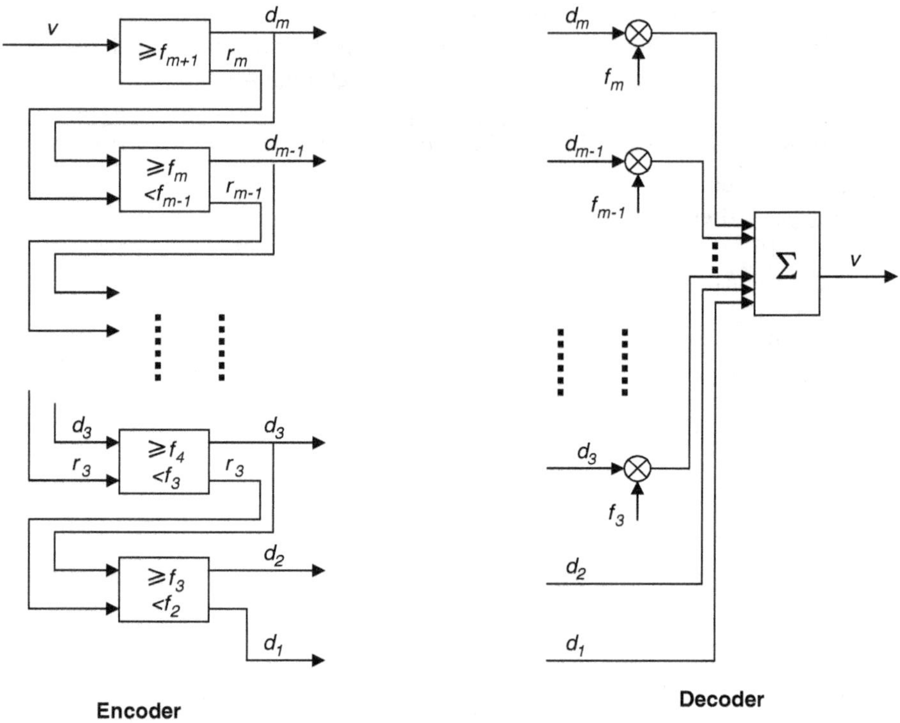

Fig. 4.4 FPF CODEC structure (based on Algorithm 4.1)

Theorem 4.3 *The procedure given in Algorithm 4.1 produces FPF Fibonacci vectors.*

Proof We first recognize that if $d_k = d_{k+1}$, no forbidden pattern will be produced regardless of the value of d_{k-1}. Therefore we only need to show that when $d_k \neq d_{k+1}$, d_{k-1} will satisfy $d_{k-1} = d_k$. Based on Algorithm 4.1, $d_k \neq d_{k+1}$ occurs only when d_k is *forced* to be 0 or 1. The proof reduces to showing that when d_k is forced to one particular value, d_{k-1} will be coded using the same value as d_k.

- If $d_k = 1$ and $d_k \neq d_{k+1}$ [d_k **in force-1 zone**]
  $\Rightarrow r_{k+1} \geq f_{k+1}$ and $r_k = r_{k+1} - f_k$
  $\Rightarrow r_k \geq f_{k+1} - f_k$
  $\Rightarrow r_k \geq f_{k-1}$ [d_{k-1} **not in force-0 zone**]
  $\Rightarrow d_{k-1} = 1$
- If $d_k = 0$ and $d_k \neq d_{k+1}$ [d_k **in force-0 zone**]
  $\Rightarrow r_{k+1} < f_k$ and $r_k = r_{k+1}$
  $\Rightarrow r_k < f_k$ [d_{k-1} **not in force-1 zone**]
  $\Rightarrow d_{k-1} = 0$ □

In Table 4.2 the complete set of 6-bit codewords generated by Algorithm 4.1 are listed as CODE-1. As stated earlier, the MSB stage is different from other stages since

Table 4.2 7-bit near-optimal FPF code books

Input	CODE-1						CODE-2					
Decimal	f_6	f_5	f_4	f_3	f_2	f_1	f_6	f_5	f_4	f_3	f_2	f_1
value	8	5	3	2	1	1	8	5	3	2	1	1
20*	1	1	1	1	1	1						
19*	1	1	1	1	1	0						
18*	1	1	1	1	0	0						
17*	1	1	1	0	0	1						
16*	1	1	1	0	0	0						
15	1	1	0	0	1	1						
14	1	1	0	0	0	1						
13	1	1	0	0	0	0						
12	0	1	1	1	1	1	1	0	0	1	1	1
11	0	1	1	1	1	0	1	0	0	1	1	0
10	0	1	1	1	0	0	1	0	0	0	1	1
9	0	1	1	0	0	1	1	0	0	0	0	1
8	0	1	1	0	0	0	1	0	0	0	0	0
7	0	0	1	1	1	1						
6	0	0	1	1	1	0						
5	0	0	1	1	0	0						
4	0	0	0	1	1	1						
3	0	0	0	1	1	0						
2	0	0	0	0	1	1						
1	0	0	0	0	0	1						
0	0	0	0	0	0	0						

there is no preceding bit and, for the values in the gray zone, d_{m-1} can be coded to be either value. In Algorithm 4.1, we arbitrarily choose to encode the MSB (d_m) to be 0 when the input value is in the gray zone. If we encode d_m to be 1 for all values in the gray zone, we end up with a different set of codewords (listed as CODE-2 in the table). All codewords in both CODE-1 and CODE-2 are FPF. For clarity, we only list codewords in CODE-2 that are different from codewords in CODE-1.

Based on Eq. 4.6, we can easily see that the total numbers of codewords in both CODE 1 and CODE-2 are f_{m+2}, slightly smaller than the maximum cardinality of $2 \cdot f_{m+1}$ given in Eq. 3.6. Since $f_{m+2} < 2 \cdot f_{m+1} < f_{m+3}$, we know that for a given input vector size n, the number of bits needed for the proposed CODEC is no more than 1 bit more than the minimum number of bits required, m_{opt}. Table 4.3 lists the number of bits needed to encode the binary data from 3 to 32 bits: n_{in} denotes the number of bits for the input binary bus, m_{opt} is the number of bits required for the optimal code, m_{no} is the number of bits needed for the proposed CODEC and $\Delta(m)$ is the difference between the two.

4.4.3 Optimal CODEC

A quick examination of Table 4.2 shows that ALL the valid FPF-CAC codewords are actually listed in the table: there are a total of $(f_{m+1} - f_m)$ codewords in CODE-2 that

Table 4.3 Overhead comparison of FPF CODECs

n_{in}	m_{opt}	m_{no}	$\Delta(m)$	n_{in}	m_{opt}	m_{no}	$\Delta(m)$
3	4	4	0	18	26	26	0
4	5	6	1	19	27	28	1
5	7	7	0	20	29	29	0
6	8	9	1	21	30	30	0
7	10	10	0	22	31	32	1
8	11	12	1	23	33	33	0
9	13	13	0	24	34	35	1
10	14	15	1	25	36	36	0
11	16	16	0	26	37	38	1
12	17	17	0	27	39	39	0
13	18	19	1	28	40	41	1
14	20	20	0	29	42	42	0
15	21	22	1	30	43	43	0
16	23	23	0	31	44	45	1
17	23	23	1	32	46	46	0

are not included in CODE-1. The total number of codewords in CODE-1 is f_{m+2}. Therefore the total number of distinct codewords in both CODE-1 and CODE-2 is:

$$\begin{aligned} T_g(m) &= f_{m+2} + f_{m+1} - f_m \\ &= f_{m+1} + f_m + f_{m+1} - f_m \\ &= 2 \cdot f_{m+1} \end{aligned} \tag{4.7}$$

We can see the reason that the near-optimal codes do not reach the maximum cardinality is that there are redundant FPF Fibonacci vectors for the values in the *gray zone*. For a coding scheme to reach the optimal overhead performance, we need to remove this redundancy.

Table 4.4 shows how such a modification is done. We simply move the codewords in the CODE-2 gray zone to the top of CODE-1. In doing so, the values these codewords represent are shifted by $f_{m+2} - f_m = f_{m+1}$. The MSB stage of the CODEC is modified to reflect this value shift in the new mapping scheme:

$$\begin{cases} d_m = 1, r_m = v - f_{m+2} & \text{if } v \geq f_{m+2}, \\ d_m = 1, r_m = v - f_m & \text{if } f_{m+2} > v \geq f_{m+1}, \\ d_m = 0, r_m = v & \text{if } v < f_{m+1} \end{cases} \tag{4.8}$$

The decoder is modified accordingly:

$$v = \begin{cases} \sum_{k=0}^{m-1} d_k \cdot f_k + f_{m+1} & \text{if } b_m b_{m-1} = 10, \\ \sum_{k=0}^{m-1} d_k \cdot f_k & \text{otherwise} \end{cases} \tag{4.9}$$

Table 4.4 gives the codewords using in the optimal CODEC, along with the value that each codeword represents. Another way to view the mapping scheme is to consider

Table 4.4 Code book for optimal FPF CODEC

Input Decimal value	XB f_7 13	OPT ENC f_6 8	f_5 5	f_4 3	f_3 2	f_2 1	f_1 1
25*	1	1	0	0	1	1	1
24*	1	1	0	0	1	1	0
23*	1	1	0	0	0	1	1
22*	1	1	0	0	0	0	1
21*	1	1	0	0	0	0	0
20*	0	1	1	1	1	1	1
19*	0	1	1	1	1	1	0
18*	0	1	1	1	1	0	0
17*	0	1	1	1	0	0	1
16*	0	1	1	1	0	0	0
15	0	1	1	0	0	1	1
14	0	1	1	0	0	0	1
13	0	1	1	0	0	0	0
12	0	0	1	1	1	1	1
11	0	0	1	1	1	1	0
10	0	0	1	1	1	0	0
9	0	0	1	1	0	0	1
8	0	0	1	1	0	0	0
7	0	0	0	1	1	1	1
6	0	0	0	1	1	1	0
5	0	0	0	1	1	0	0
4	0	0	0	0	1	1	1
3	0	0	0	0	1	1	0
2	0	0	0	0	0	1	1
1	0	0	0	0	0	0	1
0	0	0	0	0	0	0	0

the codewords (as having an extra bit d_{m+1} as shown in the second column (XB) in the table). This bit is not transmitted on the bus since its value can be recovered by the decoder, based on the values of d_m and d_{m-1}.

The overhead performance of the optimal coding scheme reaches the theoretical lower bound given in Eq. 3.10. Compared to the CODEC with near-optimal overhead performance, the optimal CODEC has added complexity. Using the optimal CODEC does not offer additional gain in overhead in the example given here, since the total number of distinct codewords increases from 21 to 26. As shown in Table 4.3, only when $f_m + 2 < 2^n < 2 \cdot f_{m+1}$ does the optimal coding offer saving of one bit. Designers should evaluate the saving against the added complexity before deciding which CODEC to use.

4.4.4 Implementation and Experimental Results

The encoder and decoder based on Algorithm 4.1 can be implemented using the structure illustrated in Fig. 4.4. The encoder converts a n-bit binary vector $v = b_n \cdots b_1$

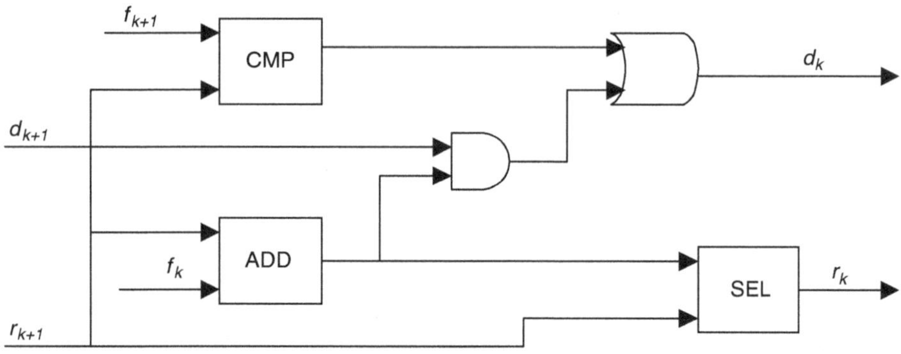

Fig. 4.5 Internal logic of the kth block in the FPF encoder

to an m-bit vector $V = d_m d_{m-1} \cdots d_1$. The m-to-n-bit decoder recovers v from the received data V.

The encoder consists of m stages. Figure 4.5 depicts the circuit details of the kth stage, where $k < m$. The inputs of the stage are outputs from the previous stage: d_{k+1} and r_{k+1}. and the outputs are d_k and r_k. Each stage consists of 1 subtraction operation, 1 comparison and a 2-to-1 multiplexer.

For the near optimal encoder, the MSB stage is simpler than the other stages, as discussed earlier. For CODE-2, the MSB stage requires only one subtractor and one selector. The MSB stage of the near-optimal CODEC can be modified to further simplify the logic. Let $b_n b_{n-1} \cdots b_1$ be the binary input vector. If we let $d_m = b_n$ and r_m be $b_{n-1} b_{n-2} \cdots b_1$, the mathematical expression of this mapping is

$$v = b_n \cdot 2^{n-1} + \sum_{k=1}^{m-1} d_k \cdot f_k \tag{4.10}$$

This modification is to simply encode the output MSB bit to the input MSB bit. The outputs are still FPF codes because to encode an n-bit binary code to an m-bit Fibonacci code, n and m satisfy $2^n < f_{m+2}$ and we have

$$2^n \leq f_{m+2} < 2 \cdot f_{m+1} \Longrightarrow 2^{n-1} < f_{m+1} \tag{4.11}$$

Therefore the n-bit input binary data can be broken into the MSB bit and an $(n-1)$ bit vector. We simply construct an encoder for the $n-1$ bit vector. The MSB bit effectively controls the output bit value when the $(n-1)$ bit input value is in the *gray zone*.

Figure 4.6 shows the modified CODEC with the simplified MSB stage. On the encoder side, the MSB of the input b_n is mapped directly to the MSB of the output d_m. The rest of the input vector $b_{n-1} b_{n-2} \cdots b_1$ becomes the input of the $(m-1)$th stage. On the decoder side, the first input of the summation stage is $d_m \cdot 2^{n-1}$, instead of $d_m \cdot f_m$ as in Fig. 4.4.

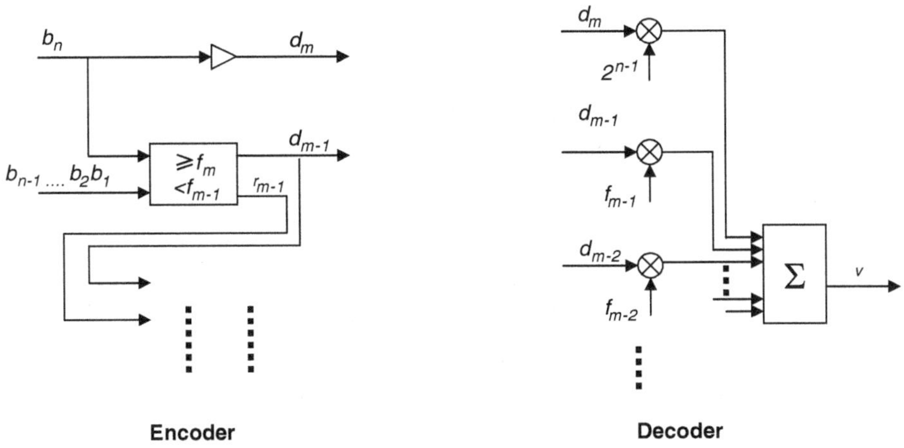

Fig. 4.6 An FPF CODEC structure with optimized MSB stage

To evaluate the complexity of the CODECs, we implemented the near-optimal CODEC in both an FPGA (Xilinx XC4VLX15-12 [9]) and an ASIC (both a TSMC 90 nm process [8] and a TSMC 130 nm process) design style. Figure 4.7 plots the resource utilization and delay of the FPGA implementation, and Fig. 4.8 plots the gate counts and delay of the encoder for input bus widths from 8 to 32, implemented

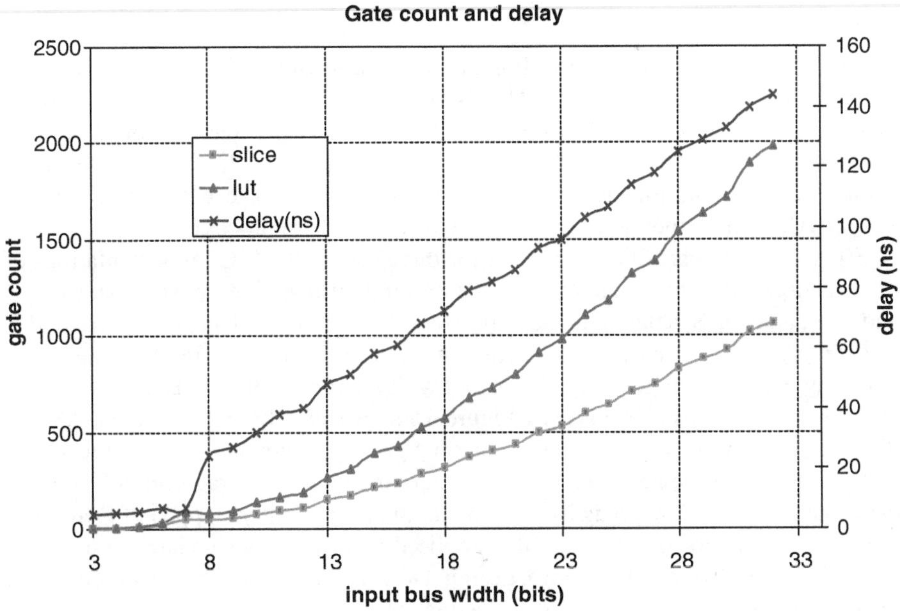

Fig. 4.7 Resource utilization and delay of the FPF encoders for different input bus sizes in FPGA implementation

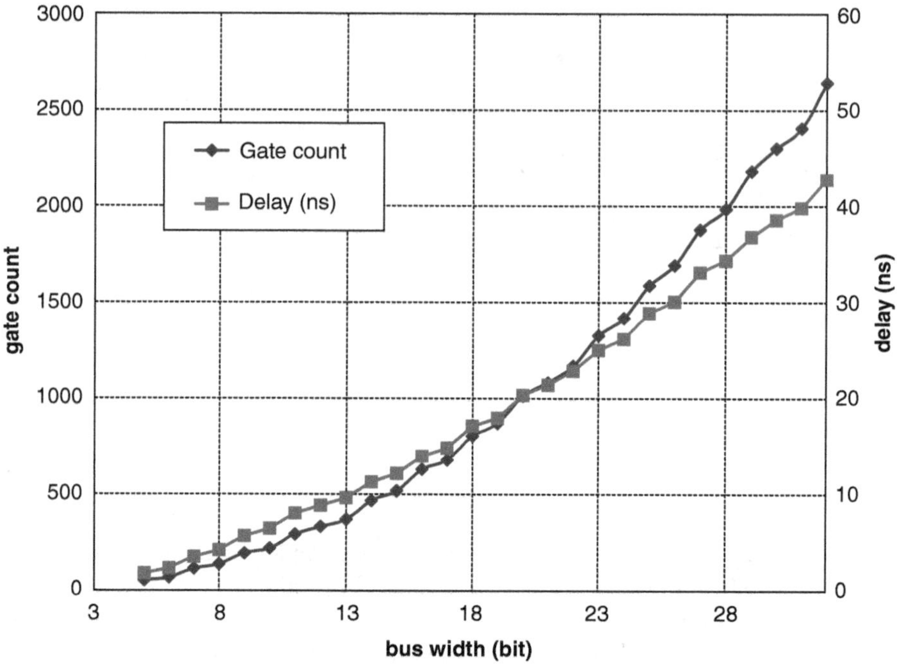

Fig. 4.8 Gate count and delay of FPF encoders for different input bus sizes in TSMC 90 nm ASIC implementation

using a TSMC 90 nm process [8]. For an input data width of 12-bits, the equivalent number of 2-input gates is 369. This is roughly two orders of magnitude lower than the gate count reported in Fig. 4.1 [89]. For the input data width of 32-bits, the total area is 17865 μm^2 and the equivalent gate count is 2,762. The sizes grow approximately quadratically with bus size, as expected. The gate counts for the 130 nm ASIC implementation are very close to the 90 nm process.

Figure 4.9 compares the gate count for three different ASIC implementations of the encoder: a flat implementation based on random mapping between data words and FPF codewords, a flat implementation based on FNS-based mapping and a multi-stage implementation based on Algorithm 4.1. All the designs are implemented in an ASIC style, using automatic synthesis tool with identical settings. The same TSMC 90 nm process was used in all cases. Figure 4.9 shows that the size of two flat designs ith random mapping and FNS-based mapping grow exponentially. The figure shows that the mapping schemes affect the encoder complexity: on average, an FPF encoder using the FNS-based mapping is 50% smaller than a randomly mapped FPF encoder. It also shows that for small busses, the proposed multi-stage design have similar gate counts compared with a single-level design. However, for busses with 8 or more bits, the proposed design offers significant savings in terms of gate count.

Figures 4.7 and 4.8 also show the delays of the encoder. Understandably, the delay increases as the input bus size increases, since the total delay is the accumulated delay

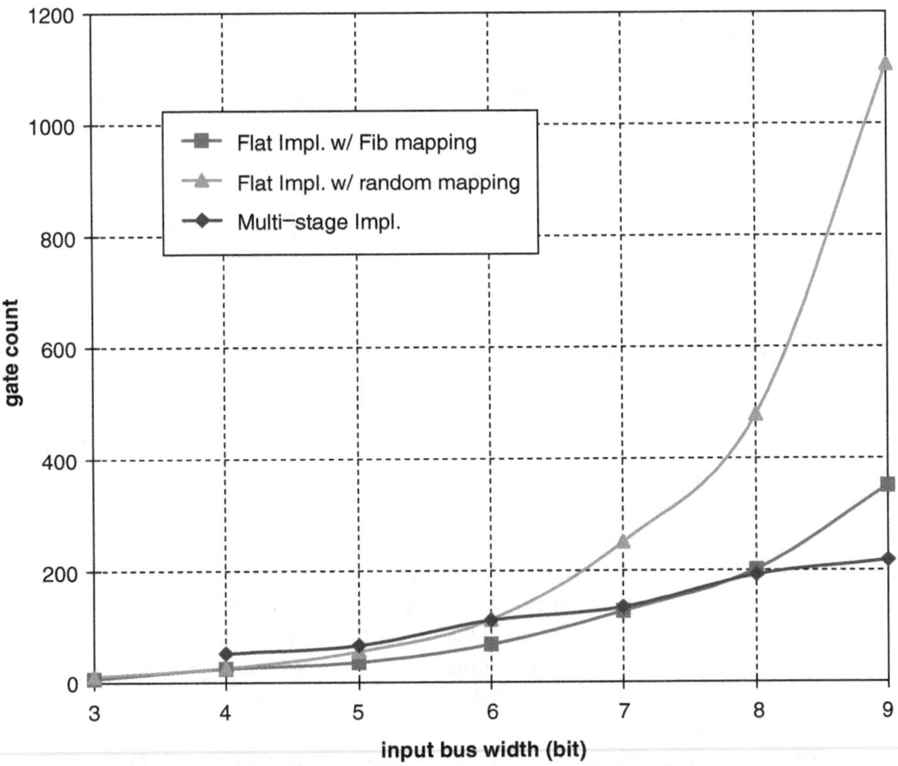

Fig. 4.9 Gate count comparison for different FPF encoder implementations in a TSMC 90 nm ASIC process: a random mapping based flat implementation, an FNS-based flat implementation and an FNS-based multi-stage implementation

of all stages. The stages near the MSB are generally slower and bigger because of the larger number of bits in both the input and the remainder. Unlike the single-level implementations, our design allows pipeline stages to be inserted between stages to mitigate this problem.

Bus partitioning can also be used to reduce the total size of the encoder, and improve the speed of both the encoder and the decoder. Our experiments confirmed that the maximum input-to-output delay of a non-pipelined m-bit encoder is $\tau(m) \propto O(m^2)$. Partitioning the bus into two groups can improve the bus speed by approximately a factor of 4. Similarly, the total area also has an approximately quadratic growth with the number of input bits, and therefore partitioning the bus reduces the total area by $\sim$50%.

The decoder structure is simpler than the encoder, and incurs no ripple delay. However, as the bus width grows, the adder size increases and more delay will be incurred. Note that there is no multiplication or AND operation required in the decoder implementation. Since f_k is a constant, we simply need to connect d_k to the non-zero bit positions of f_k and then perform the final addition using a faster adder.

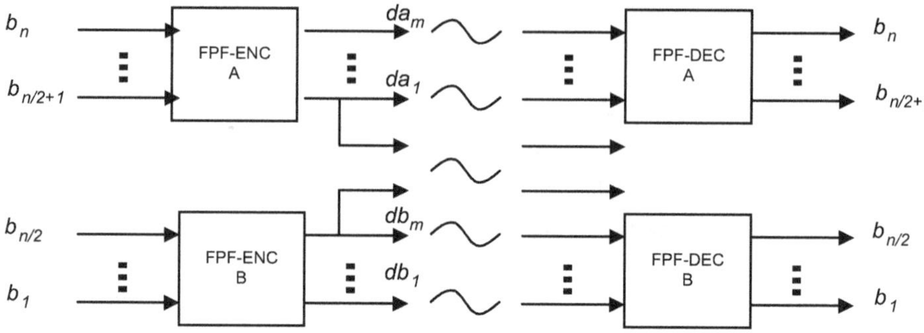

Fig. 4.10 FPF CODEC block diagram with bus partition for delay/area improvement

Figure 4.10 illustrates a structure where an n-bit input bus is split into two $\frac{n}{2}$ -bit groups. Each group is encoded and decoded independently. The maximum delay of the encoder and decoder stages are $\tau_{enc}(n/2)$ and $\tau_{dec}(n/2)$, instead of $\tau_{enc}(n)$ and $\tau_{dec}(n)$. The gate count reduces to $2 \cdot GC(n/2) \cong \frac{1}{2}GC(n)$.

The crosstalk that occurs across the group boundaries must be dealt with when bus partitioning is employed. In Fig. 4.10, we simply duplicate the boundary lines. This requires two extra wires for each added partition. Other techniques discussed in earlier sections such as "group complement" are applicable as well. The trade-off between area and speed can be balanced to achieve the required speed, while minimizing the additional area overhead.

The FPF code was originally proposed to improve the bus speed. However, our simulations show that by applying FPF encoding, bus energy consumption can be reduced as well. To demonstrate this, we randomly generated 10,000 input vectors for 8, 12, 16 and 32 bit busses and transmitted these vectors on an uncoded bus, and a coded bus respectively. For each bus width, the normalized total energy consumption, E_{norm} is computed based on Eq. 6.7 with V_{dd} and C_L both set to 1. Table 4.5 provides the normalized energy consumption for the coded and uncoded busses of several different sizes. The comparison is technology independent and is also independent of bus configurations (i.e., bus length, wire sizing, wire material and layer etc., provided that $\lambda \gg 1$ is guaranteed). The results indicate that even with ~44% more wires, coded busses have a lower total energy consumption. It is important to point out that these savings are achieved using random sequences. For busses that transmit correlated data, results may vary. Note that the simulations in Table 4.5 do not include the power consumption of the CODEC.

Table 4.5 Power consumption comparison between coded and uncoded busses

Input bus size	E_{norm} uncoded	E_{norm} coded	Saving
8	35075	27825	20.7 %
12	55004	44222	19.6 %
16	74999	61456	18.6 %
32	154191	124148	19.5 %

4.5 FTF-CAC CODEC Design

In this section, we will show that the FNS-based mapping can be used in FTF CODEC designs as well. Even though we have proven in Sect. 3.1.2 that the FTF codes have a slightly higher theoretical lower bound of bus overhead than FPF codes, we will show in this section that FTF CODECs can be implemented more efficiently than FPF CODECs. They are also easier to implement in a partitioned bus.

4.5.1 Mapping Scheme

Similar to the FPF CODEC design, the FTF CODEC design is also based on mapping the input into the Fibonacci Numeral System. We first show that such a mapping exists.

Theorem 4.4 *$\forall v \in [0, f_{m+2} - 1]$, $\exists d_m d_{m-1} \cdots d_2 d_1$, S.T. $v = \sum_{k=1}^{m} d_k \cdot f_k$ and $d_m d_{m-1} \cdots d_2 d_1$ is FTF.*

Theorem 4.4 states that for any number $v \in [0, f_{m+2})$, there exists at least one m-bit Fibonacci vector $d_m d_{m-1} \cdots d_2 d_1 = v$ which represents v in FNS and also satisfies the *forbidden transition free* property.[1] In other words, any number in the range of $[0, f_{m+2})$ can be mapped to an m bit FTF codeword in the FNS space.

Proof Let us first define S_{00} as the set of all k-bit Fibonacci vectors with the two MSBs of "00". Consider a Fibonacci vector $V_{00} \in S_{00}$, we know $V_{00} \in [0, f_k)$ based on Eq. 4.6. Similarly, we define sets S_{01}, S_{10} and S_{11}, with representative members V_{01}, V_{10} and V_{11} respectively. We have that $V_{01} \in [f_{k-1}, f_{k+1})$, $V_{10} \in [f_k, 2f_k)$ and $V_{11} \in [f_{k+1}, f_{k+2})$. This is shown graphically in Fig. 4.11. We can see from Fig. 4.11 that the ranges of these four sets of vectors overlap. For any vector $v \in S_{01}$, we can

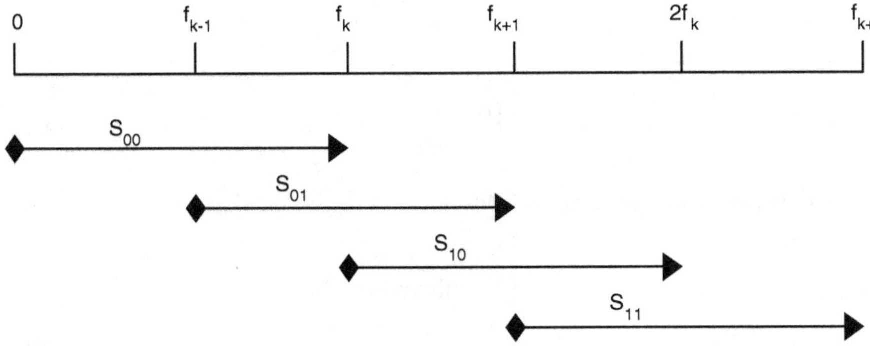

Fig. 4.11 Range of the vectors with MSBs of "00", "01", "10" and "11"

[1] The FTF property requires that 01 is prohibited at $b_{2k-1}b_{2k}$ boundaries and 10 at $b_{2k}b_{2k+1}$ boundaries.

find an equivalent vector in S_{00} or S_{10}. Similarly, any vector $v \in S_{10}$ can be replaced by an equivalent vector in S_{01} or S_{11}.

Now for a given number v, we can find an m bit vector $V_{in} = d_m d_{m-1} \cdots d_2 d_1$ in the FNS which represents v (based on the completeness of FNS). Now assuming that "01" is the prohibited pattern at the $d_k d_{k-1}$ boundary, we can replace $d_k d_{k-1} \cdots d_1$ with an equivalent k-bit vector in V_{00} or V_{10} to produce a new m bit vector V_{eq} which is equivalent to V_{in}. Note that V_{eq} has no prohibited pattern at $d_k d_{k-1}$ boundary.

Also since "01" is the prohibited pattern at $d_k d_{k-1}$, the prohibited pattern at boundaries $d_{k+1} d_k$ and $d_{k-1} d_{k-2}$ is "10". We know that

- If V_{in} does not have the prohibited pattern at $d_{k+1} d_k$, then V_{eq} does not introduce the prohibited pattern at $d_{k+1} d_k$, since the above manipulation causes the kth bit value to either change from '0' to '1' or no change at all.
- V_{eq} does not have the prohibited pattern at $d_{k-1} d_{k-2}$ since d_{k-1} in V_{eq} is '0'.

The first condition guarantees that the manipulation do not introduce a prohibited pattern into more significant boundaries. The second condition guarantees that the manipulation does not introduce prohibited patterns in the next less significant boundary.

Therefore, by replacing the prohibited pattern iteratively from the MSB to the LSB, the final result is a vector which is equivalent to V_{in}, and is prohibited pattern free (and therefore it is a FTF codeword). This proves Theorem 4.4. □

4.5.2 Coding Algorithm

To design a coding algorithm that maps each input value to an FTF codeword in the FNS, we observe that in Fig. 4.11, if d_k is coded as

$$d_k = \begin{cases} 0 & \text{if } v < f_k, \\ 1 & \text{otherwise} \end{cases} \tag{4.12}$$

and d_{k-1} is coded as

$$d_{k-1} = \begin{cases} 0 & \text{if } v < f_{k+1}, \\ 1 & \text{otherwise} \end{cases} \tag{4.13}$$

then "01" is eliminated on $d_k d_{k-1}$. Conversely, if d_k, is coded as

$$d_k = \begin{cases} 0 & \text{if } v < f_{k+1}, \\ 1 & \text{otherwise} \end{cases}$$

and d_{k-1} coded as

$$d_{k-1} = \begin{cases} 0 & \text{if } v < f_{k-1}, \\ 1 & \text{otherwise} \end{cases}$$

then "10" is eliminated on $d_k d_{k-1}$.

Based on the above observations, the coding algorithm is shown in Algorithm 4.2. The algorithm generates FTF codewords with "01" prohibited at $d_{2k+1}d_{2k}$ and "10" prohibited at $d_{2k}d_{2k-1}$.

Algorithm 4.2 FTF encoding algorithm: FTF-CAC(v)

$r_{m+1} = v$;
for $k = m$ downto 2 **do**
  if $r_{k+1} < f_{2\lfloor \frac{k}{2} \rfloor + 1}$ **then**
    $d_k = 0$;
  else
    $d_k = 1$;
  end if
  $r_k = r_{k+1} - f_k \cdot d_k$;
end for
$d_1 = r_2$;
Return: $d_m d_{m-1} \cdots d_1$;

$\lfloor \cdot \rfloor$: floor operator.

The encoding is carried out sequentially in $m - 1$ stages, similar to a division operation. Starting from the MSB, each stage compares the input to a Fibonacci number and produces a coded bit as well as a remainder. The remainder from one stage becomes the input to the next stage.

4.5.3 *Implementation and Experimental Results*

A direct implementation of the encoder and decoder based on Algorithm 4.2 is illustrated in Fig. 4.12 (here we assume m is even). The encoder, which converts a n-bit binary vector $v = b_n \cdots b_1$ to an m-bit Fibonacci vector $V = d_m d_{m-1 \cdots d_1}$, consists of $m - 1$ stages. The decoder converts the codewords back to the original n-bit binary vector.

The decoder in Fig. 4.12 is exactly the same as the decoder for FPF code shown in Fig. 4.12. It consists of an m input adder. The m inputs of the adder have different numbers of bits with the most significant input having the most number of bits ($\lceil \log(f_m) \rceil$) and two least significant inputs having 1 bit each. The m input adder is a special case of an arithmetic sum of product block, and can be synthesized efficiently [33]. The overall complexity is approximately $O(m^2/2)$.

$$d_k = \begin{cases} 0 & \text{if } r_{k+1} < f_{2\lfloor \frac{k}{2} \rfloor + 1}, \\ 1 & \text{otherwise} \end{cases} \tag{4.14}$$

$$r_k = r_{k+1} - f_k \cdot d_k \tag{4.15}$$

The encoder stages implement the arithmetic function given in Eq. 4.15. Each stage produces a single bit d_k and a remainder r_k. For the MSB stage, the input r_{k+1} is the

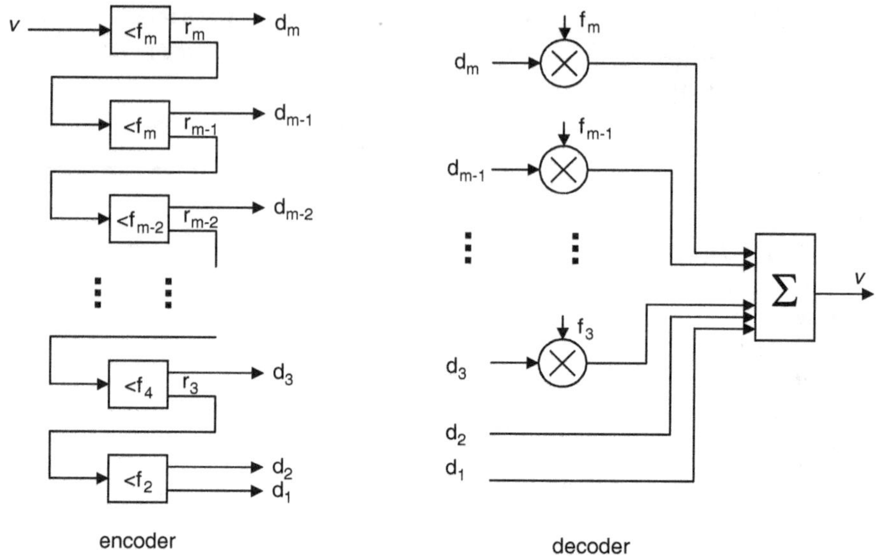

Fig. 4.12 FTF encoder and decoder block diagram

input data (v) to be encoded. For other stages, r_{k+1} is the remainder of the preceding stage. If m is even, the implementation requires *one* comparator, *one* adder and *one* 2:1 MUX. If m is odd, the implementation only needs one adder and one MUX since the adder (subtraction) also functions as a comparator. Assuming the hardware resources required for a comparator are the same as for an adder, the total resources required for an m bit bus encoder is $\frac{3}{2}m$ adders and m MUXes. The number of bits reduces monotonically from the MSB stage to the LSB stage.

If we combine every two stages in the encoder carefully, the logic can be further simplified. Let "10" be the prohibited pattern for $d_k d_{k-1}$ boundaries. A two-bit encoder stage implements the following:

$$d_k d_{k-1} = \begin{cases} 00 & \text{if } r_{k+1} < f_k, \\ 01 & \text{if } r_{k+1} < f_{k+1}, \\ 11 & \text{otherwise} \end{cases}$$

$$r_{k-1} = \begin{cases} r_{k+1} & \text{if } r_{k+1} < f_k, \\ r_{k+1} - f_k & \text{if } r_{k+1} < f_{k+1}, \\ r_{k+1} - f_{k+1} & \text{otherwise} \end{cases} \tag{4.16}$$

Equation 4.16 can be implemented using *two* adders and *two* MUXes. Since the single-bit stage implementation requires the resources for *three* adders and *two* MUXes to encode two bits, the two-bit stage implementation of Eq. 4.16 is simpler. We should point out that this simplification is only achieved when "10" is the prohibited pattern. We can not reduce the implementation complexity if "01" is the

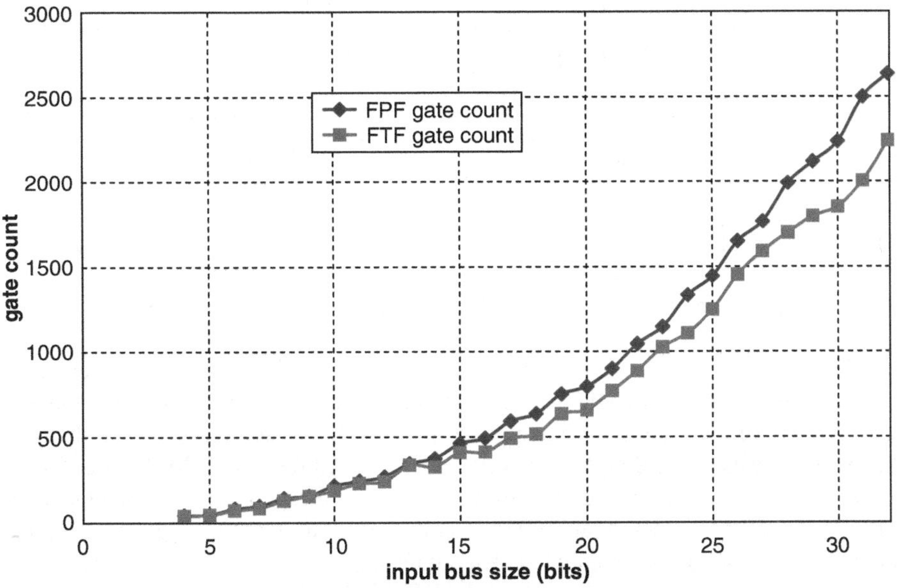

Fig. 4.13 FTF and FPF encoder gate count comparison

prohibited pattern in the two bit implementation. For comparison, the FPF CODEC described in Sect. 4.4 need one adder, one comparator and one MUX for each stage, nearly twice the complexity of the FTF CODEC with the two-bit combined implementation. The FTF and FPF decoders are identical.

To evaluate the complexity of the CODECs, we synthesized the CODEC in a 90 nm process [8] using an ASIC design approach. Figure 4.8 plots the equivalent gate counts of the FTF encoder for input bus widths from 4 to 32. For a 12-bit input data width, the equivalent number of 2-input gates for the FTF implementation is 245. For the input data width of 32-bits, the equivalent gate count for the FTF encoder is 2247. The growth of the FTF encoder size is quadratic with respect to the bus size, similar to the FPF CODECs.

For comparison, the gate counts of the ASIC implementation of FPF-CAC encoders obtained through the same synthesis flow are also plotted in Fig. 4.13. The FPF-CAC encoder gate count for a 32-bit bus is 2640,[2] which is 17.6% greater than the FTF-CAC encoder gate count for the same bus size. On average, the gate count for the FPF encoder is ~17% higher that the FTF-CAC encoder. The decoders for both codes are identical.

Without pipelining, the overall delay of the encoder is the summation of all the stage delays. This total delay can be be significant. Fortunately, our design allows

[2] This is slightly different from the number given in Sect. 4.4.4 due to different versions of the synthesis tool used in experiments.

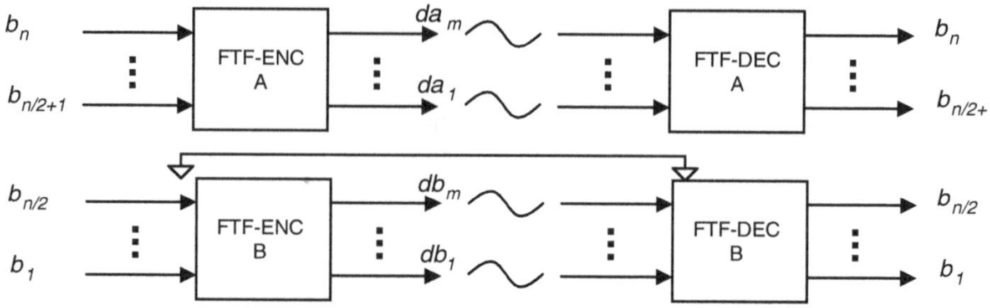

Fig. 4.14 FTF CODECs combined with bus partitioning

pipeline stages to be inserted between stages. The speed of the encoder is determined by the slowest stage (the MSB stage). It is reported in [17] that a 64-bit adder has a total delay of less than 250 ps using a 65 nm process. With the additional delay of the MUX, we estimate that the slowest stage of our CODEC (the MSB stage) has a delay of no more than 300 ps. Therefore, we expect a pipelined implementation of the FTF CODEC to operate at 3 GHz. It is important to note that pipelining cannot be used in an encoder implemented as a flat design.

Again, the complexity and speed of the FTF CODEC can be further improved by applying bus partitioning. The maximum input-to-output delay of the non-pipelined m-bit encoder is $\tau(m) \propto O(m^2)$. The total area also exhibit a quadratic growth with the number of input bits. Therefore partitioning the bus will reduce the total area by $\sim$50%. Unlike FPF CODECs, which require either two shield wires between the group boundaries or more complicated techniques such as group complement for bus partitioning, the FTF code only needs one grounded wire between two adjacent groups (as shown in Fig. 4.14). Therefore, the slight bus area overhead advantage of FPF no longer exists when bus partitioning is applied.

4.6 Summary

This chapter covers CODEC designs for memoryless CACs. In particular, the discussion was restricted to $3C$-free codes. We confirmed that a flat implementation of a random or FNS-based mapping is impractical for large busses because of the exponential growth of CODEC size and the rapid speed degradation associated with the size growth.

A challenging task in CAC code design is to find an efficient mapping. A quick analysis shows that even a simple $4 \Rightarrow 5$-bit CODEC will have over $2 \cdot 10^{13}$ possible mappings, hence exhaustive search is impossible.

To address this, we presented several solutions. We first presented two bus partitioning techniques, "group complement" in Sect. 4.2 and "bit-overlapping" in Sect. 4.3. By partitioning the bus, the encoders and decoders are guaranteed to have limited complexity, even if the mapping used are suboptimal, the associated penalty is acceptable due to the reduced size of the bus. Due to the lack of an optimal mapping scheme, we implemented partitioned CODECs by using an arbitrary mapping. The undesired crosstalk patterns across the group boundaries are dealt with by complementing the group outputs or overlapping input bits. Even though the CODEC sizes are small, both schemes incur an extra area overhead and therefore are not deemed efficient.

We then presented solutions to the mapping which use the Fibonacci numeral system (FNS). Using the FNS based mapping solutions, we developed several CODECs for both FPF and FTF $3C$-free codes in Sect. 4.4 and 4.5 respectively.

For FPF codes, we first presented the mathematical proof that any number v has at least one corresponding FPF vector in the FNS. Based on this mapping, we developed an encoding algorithm that produces FPF codes with near-optimal cardinality (i.e., the coding scheme yields an area overhead which is at worst one bit greater than the theoretical lower bound). We then presented an improved FPF coding scheme, which achieves the theoretical bound in overhead efficiency. We showed that the optimal FPF encoding also follows the FNS mapping.

We also discussed FPF CODEC implementations. Encoders of both FPF CODECs can be implemented in multi-stage structure. Further optimizations were proposed to eliminate the MSB stage in the encoder, thereby simplifying the decoder as well. We implemented the near-optimal FPF CODEC in both an FPGA as well as a 90 nm ASIC flow. The reported results show a quadratic growth in both CODEC size and delay, with the increasing bus size. Our FNS-based CODECs are all several orders of magnitude smaller than a previously reported flat random mapping based CODEC [89]. The equivalent 2-input gate count for our FPF encoder is about 4,000 for a 32-bit bus. We also presented techniques to combine our approach with bus partitioning to address the propagation delay issue as well as to reduce the total size of the CODEC.

We also compared the average bus energy consumption of uncoded and FPF coded busses, through simulation. Our experimental results show that FPF coding offers on average ~20% energy saving.

We also presented FNS-based FTF CODEC designs. The existence of an FTF mapping in FNS for any number was first proven mathematically. We showed that FNS-based FTF encoding algorithms are very efficient in terms of area overhead. We showed that the main advantage of FTF CODEC designs lie in their efficient implementations. FTF encoders can be implemented in multi-stage structures, very similar to FPF encoders. However, each stage in an FTF encoder is simpler than the corresponding stage of the FPF encoder in implementation. By combining two stages, the overall complexity of an FTF encoder is further reduced, requiring about half the number of arithmetic operations compared to the FPF encoder of the same bus size. Our ASIC-based implementation of the FTF encoders showed over 17% gate count reduction and nearly 50% speed improvement over FPF encoders implemented in the same technology with the same design tools. Another benefit offered by the FTF encoder is that it yields more efficient partitioned implementation compared to the FPF encoder.

All the CODEC designs based on FNS mapping have the following features:

- Efficiency: Both the encoder and decoder complexities grow quadratically (as opposed to exponentially) with increasing bus size.

- Modularity: Both the encoder and decoder are constructed in a systematic fashion. The encoder consists of multiple stages, and a CODEC design for a larger bus can be generated from a CODEC of a smaller bus.
- Simplicity: Bus partitioning is elegantly handled.
- Speed: The multi-stage implementation of FNS-based encoders enables easy pipelining and extreme high-speed operation. In a non-pipelined implementation, these encoders encounter a ripple delay effect and are slow for large bus sizes.

Chapter 5
Memory-based Crosstalk Avoidance Codes

Codewords generated in memory-based crosstalk avoidance encoding satisfy a specified crosstalk upper-bound requirement *with respect to* the previous codeword on the bus. The codeword is generated based not only on the current data word, but also on the previously transmitted codeword. Such a code is therefore called "memory-based code". The decoder also uses the current received codeword and either the previous received codeword or the previous decoded data word to decode the current data word. The memory depth of CACs we consider is one.

Memoryless crosstalk avoidance codes are attractive due to the simplicity of their CODEC implementations. Memory-based codes, on the other hand, offer the advantage of lower bus area overhead as reported in [35, 99]. However, they generally require more complex CODECs.

In this chapter, we first present a memory-based 4C-free CAC design that requires simple CODEC logic and 33% bus wiring overhead. We then present two generic memory-based code design methods: the pruning-based search and the ROBDD-based search.

5.1 A 4C-Free CAC

The proposed 4C-free CAC eliminates all 4C crosstalk patterns on the bus and therefore the bus satisfies $\max_j(C_{eff,j}) = (1 + 3\lambda)C_L$. For DSM processes, the speed is improved by ~33% since $\lambda \gg 1$. According to Table 2.4, for a 20 mm bus with 60× drivers, the signal delay is reduced from over 1500 to 1068 ps, and consequently a 27% improvement is achieved in practice. (Note that this value may vary slightly for different driver sizes or bus lengths). These codes are less aggressive in reducing crosstalk compared to other CACs discussed in previous chapters. However, they incur a lower area overhead penalty, and therefore can be attractive in designs where area constraints are stringent.

5.1.1 A 4C-free Encoding Technique

To encode input data into 4C-free codewords, we split the input bus into groups of 3-bits. These group signals are referred to as *group data signals*. For each group, we

C. Duan et al., *On and Off-Chip Crosstalk Avoidance in VLSI Design*,
DOI 10.1007/978-1-4419-0947-3_5, © Springer Science+Business Media LLC 2010

add an additional signal called *group complement indicator*. When a group complement indicator toggles, the bus receiver complements all the group data signals in that group.

Let V^k be the input vector in the kth cycle. For the kth cycle, we denote the group data signals of group i as $d^k_{i,j}$ (for $j = 1,2,3$) and the group complement indicator of group i as c^k_i.

For two consecutive vectors V^k and V^{k+1}, let $d^k_{i,1} = v_1$, $d^k_{i,2} = v_2$, $d^k_{i,3} = v_3$ and $c^k_i = v_4$. Also, let $d^{k+1}_{i,1} = v_5$, $d^{k+1}_{i,2} = v_6$, $d^{k+1}_{i,3} = v_7$ and $c^{k+1}_i = v_8$. We denote the group transition as $(v_1, v_2, v_3, v_4)^k_i \rightarrow (v_5, v_6, v_7, v_8)^{k+1}_i$. If we refer only to the group data signals of group i, their transition is denoted as $(v_1, v_2, v_3)^k_i \rightarrow (v_5, v_6, v_7)^{k+1}_i$.

Whenever a $4C$ transition is detected on the group data signals, i.e., $v_1 = \overline{v_5}$, $v_2 = \overline{v_6}$, $v_3 = \overline{v_7}$ and $v_1 = v_3 = \overline{v_2}$ the group data is transmitted as $(v_1, v_2, v_3, v_4)^k_i \rightarrow (v_1, v_2, v_3, \overline{v_4})^{k+1}_i$. The receiver complements the group data signals accordingly (since the group complement indicator performs a transition) and correct data is recovered.

5.1.2 An Example

Consider two successive vectors (in the kth and k+1th cycles) on a 9-bit uncoded bus

$$(010\ 101\ 010)^k \rightarrow (101\ 010\ 101)^{k+1}.$$

Spaces are provided to indicate group partitions. Note that $4C$ transitions occurs in all the groups. Now assume the bus is encoded using the scheme given above, and also assume that the initial values of group complement indicators are 0 in the kth cycle. The transmitted 12-bit vector sequence becomes

$$(0100\ 1010\ 0100)^k \rightarrow (0101\ 1011\ 0101)^{k+1}.$$

Note that in the transmitted vector above, none of the group data signals perform a transition. In this fashion, $4C$ transitions never occur within a group.

On further reflection, we realize that forbidden transitions can occur *across* groups if this scheme is utilized. Consider a 6-bit bus and assume that the initial values of all group complement indicators are 0. The un-encoded bus transitions are

$$(010\ 101)^k \rightarrow (101\ 011)^{k+1}.$$

After encoding, the transmitted vector would be

$$(010\underline{0\ 10}10)^k \rightarrow (010\underline{1\ 01}10)^{k+1}.$$

Note that in this case, a $4C$ transition occurs across the two groups (at the underlined bit positions).

We observe that such a $4C$ transition can *only* occur when the group complement indicator transitions. More precisely, it occurs when the group complement indicator of group i performs a $0 \rightarrow 1$ (or $1 \rightarrow 0$) transition and the data signals of the group $i+1$ perform a $(10v)^k_{i+1} \rightarrow (01v)^{k+1}_{i+1}$ (or $(01v)^k_{i+1} \rightarrow (10v)^{k+1}_{i+1}$) transition, where $v \in \{0,1\}$. Therefore this condition can be avoided by modifying the original encoding scheme in the following manner:

Whenever 0 (*or* 1) $= c^k_i \neq c^{k+1}_i$, and the group data signals of group $i+1$ are of the form $(10v)^k_{i+1} \rightarrow (01v)^{k+1}_{i+1}$ (or $(01v)^k_{i+1} \rightarrow (10v)^{k+1}_{i+1}$) where $v \in \{0,1\}$, we complement c^{k+1}_{i+1} and the group data signals of group $i+1$.

This complicates the bus encoding, possibly forces a rippling of data in the group complement logic and results in a slower encoder.

Since every group of 3 input wires requires an additional wire, the area overhead of this coding scheme is

$$OH_{4C} = 33\%. \tag{5.1}$$

5.2 Codeword Generation by Pruning

Graph theory based techniques can be very useful in analyzing situations involving a set of elements in which various pairs of elements are related by certain properties [19, 95]. In the case of crosstalk-based bus encoding, a graph $G^{kC}_m = (V,E)$ can be constructed for an m-bit bus.[1] G^{kC}_m consists of a set of 2^m *vertices*, V and a set of edges, E. Each $v \in V$ represents one of 2^m codewords that may be transmitted on the bus. An edge between two vertices a and b (denoted as (a,b)) indicates that transition between these two codewords satisfies the crosstalk constraint, i.e., $\max(C_{eff,j}) \leq kC$ for the transition between a and b. The properties of G^{kC}_m include

1. $|G^{kC}_m| = 2^m$, where $|G^{kC}_m|$ is the order of the graph.
2. G^{kC}_m is undirected[2].
3. For any $v \in V$, $(v,v) \in E$

Figure 5.1 illustrates the graphs for a 3-bit bus under different crosstalk constraints.

The degree of a vertex, a, in G^{kC}_m, denoted as $d_G(a)$, is the number of edges incident with the vertex. It represents the number of *legal* transitions from the current bus vector, a. $d_G(a)$, and therefore, is the number of distinct data words that can be transmitted on the bus.

G^{kC}_m can be thought of as representing an m-bit bus. Any value can be present on the bus and therefore the number of data words that can be encoded on to this m-bit

[1] To simplify the notation, we sometimes use "kC" in place of "kC-*free*". A kC bus is $(k+1)C$-free and therefore $kC \equiv (k+1)C$-free.

[2] An edge is undirected if the initial vertex and the terminal vertex are not differentiated, i.e., $(a,b) \equiv (b,a)$. This property is warranted by the symmetry of bus crosstalk, i.e., $X(a,b) \equiv X(b,a)$.

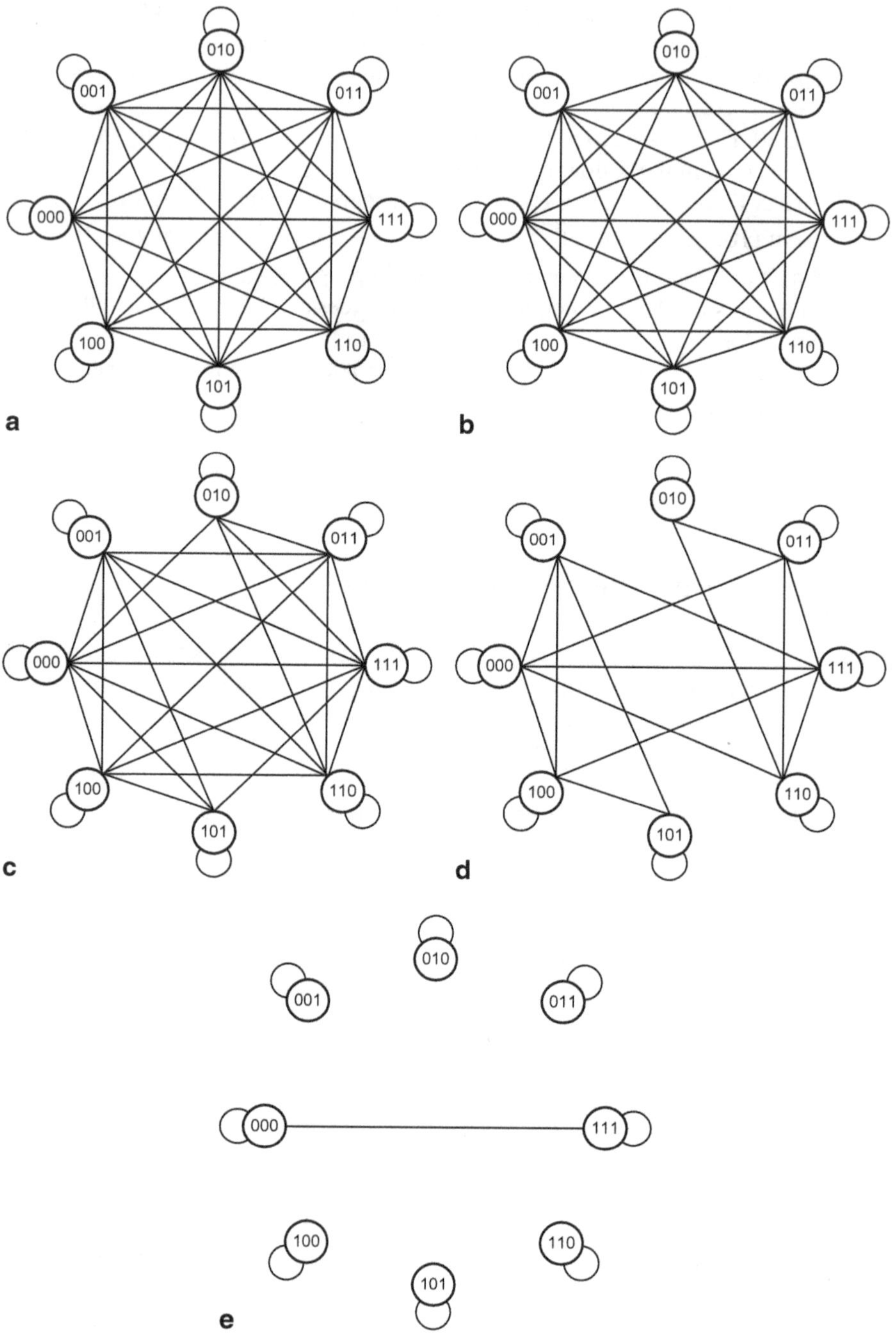

Fig. 5.1 Graphs for 3-bit busses under different crosstalk constraints **a** G_3 graph **b** $G_3^{4C-free}$ graph **c** $G_3^{3C-free}$ graph **d** $G_3^{2C-free}$ graph **e** $G_3^{1C-free}$ graph

$(k+1)C$-free bus is limited by the vertices with smallest degree. In order to transmit n-bit data, G_m^{kC} must satisfy

$$\min_{a \in V} [d_{G_m^{kC}(a)}] \geq 2^n. \tag{5.2}$$

where $d_{G_m^{kC}(a)}$ is the degree of vertex a in G_m^{kC}.

If we constrain the bus so a particular codeword is never present on the bus, it is equivalent to removing the corresponding vertex from the original graph, G_m^{kC}. Such a process is called "code pruning". Code pruning produces a vertex-induced subgraph, $G_P = (V_p, E_p)$ of G_m^{kC}. Our goal is to find such a subgraph such that $\min_{a \in V_p}(d_{GP}(a))$ is maximized. We equivalently refer to $\min_{a \in V_p}(d_G(a))$ as $\min(d_{G_p})$.

To find such a subgraph, we start with the original graph, G_m^{kC}, and determine the minimum degree of the original graph, d_{min}. We then remove all vertices with degree of d_{min} from G_m^{kC} to form an induced subgraph, G_P. The degrees of all vertices in G_p are updated and the minimum degree of the subgraph, $\min(d_{GP})$ is computed. This process of forming a new subgraph by removing the vertices with the smallest degree is iterated until the stop condition is met.

The stop condition can be set as either (a) when $\min(d_{GP})$ start decreasing, or (b) when all vertices are removed from the graph. $|G_P|$ decreases monotonically with each iteration. The value of $\min(d_{GP})$, on the other hand, does not change monotonically. The value typically increases first after each iteration before starting to decrease. Also, a local maxima may exist for $\min(d_{GP})$ and therefore search based on stop condition (a) may not return the optimum value. stop condition (b) guarantees that the global optimum is found but often results redundant iterations. A third stop condition can be used, in which additional iterations are performed after a maximum is found. The number of additional steps, however, is dependent on the size of the bus. Algorithm 5.1 gives the pseudo-code of the code pruning based on stop condition (b).

Algorithm 5.1 Iterative code pruning

```
Initial condition: G_P = G_m^{kC-free}, V_p includes all 2^m codewords;
d_opt = 0;
while V_p ≠ Ø do
   compute min(d_GP);
   for all v_i satisfies d_GP(v_i) = min(d_GP) do
      G_P = G_P − v_i;
   end for
   if min(d_GP) ≥ d_opt then
      d_opt = min(d_GP);
      G_opt = G_P;
   end if
end while
return G_opt;
```

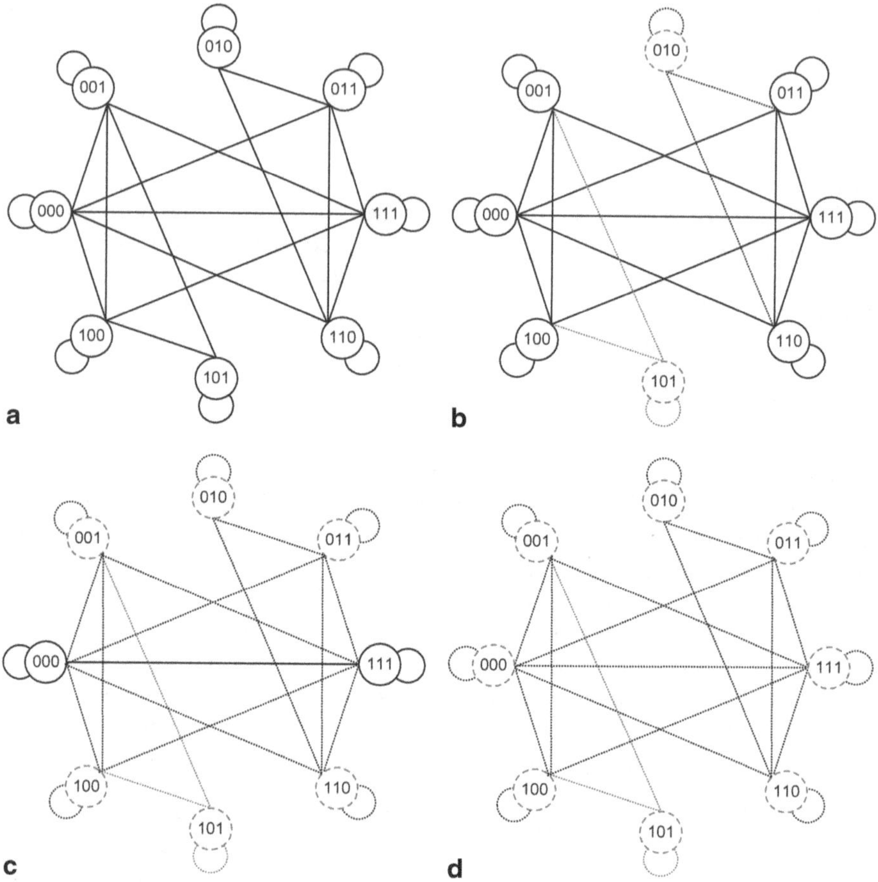

Fig. 5.2 Pruning iteration of a 3-bit $1C$ transition graph **a** In the initial graph, $\min(d_{G_3^{1C}}(v)) = 3$ and $|G| = 8$. Two vertices of degree 3 are 010 and 101. **b** In first iteration, vertices with fewest edges are removed (dash-lined circles). Edges incident to them are deleted (shown as dash lines). In the resulting subgraph G′, $\min_{v \in V'}(d_{G'}(v)) = 4$ and $|G'| = 6$ **c** In the second iteration. 4 vertices with fewest edges are removed. $\min_{v \in V'}(d_{G'}(v)) = 2$ and $|G'| = 2$ **d** After the final iteration, the resulting subgraph is an empty graph (all vertices are removed). $\min_{v \in V'}(d_{G'}(v)) = 0$, $|G'| = 0$

Figure 5.2 illustrates the pruning process for a 3-bit, $2C$-free code. In the initial graph, there are 8 vertices with $1C$ edges. Two of the vertices 101 and 010, have 3 edges each, while other vertices have larger degrees. By removing the two vertices with minimum degrees, we obtain a pruned graph G' with 6 vertices {000, 001, 011, 100, 110, 111} and the degree of the new graph is 4. The new graph has a higher minimum degree. Four vertices, {001, 011, 100, 110}, in G′ are with the lowest degree of 4. Removing these vertices results in a new graph with only two vertices 000 and 111. The degree is 2 for this graph. One more iteration will remove both

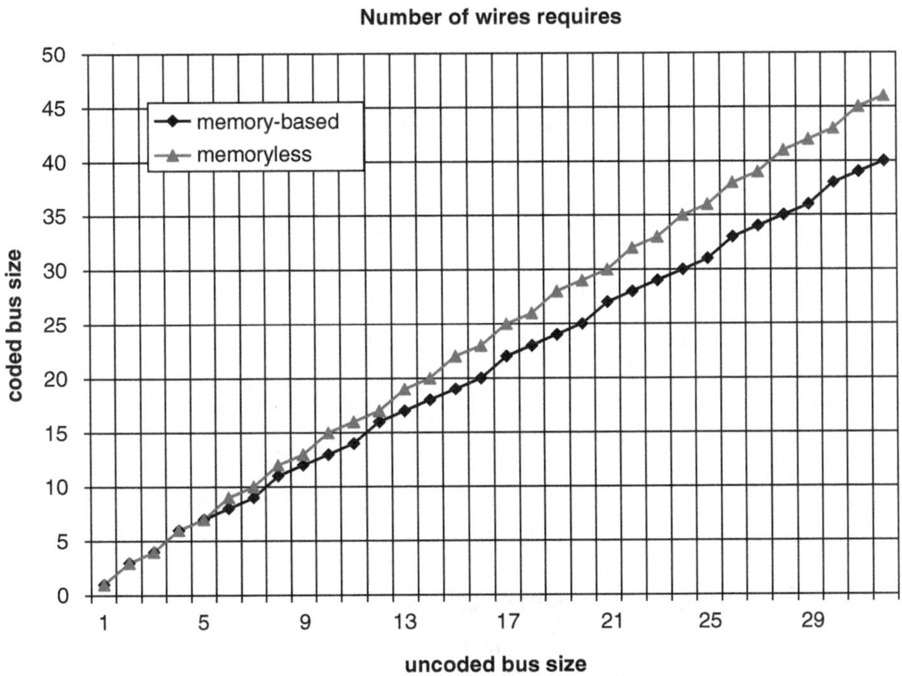

Fig. 5.3 Area overhead comparision between FTF codes and memory-based codes generated with class pruning

remaining vertices and results in an empty set. Among all the intermediate subgraphs, the subgraph after the first iteration (depicted in Fig. 5.2b) with vertices {000, 001, 011, 100, 110, 111} is a $1C$ graph that has maximum minimum degree. Note that the resulting graph is not a clique.

We note that the code pruning also applies to memoryless codes generation. For the codes to be memoryless, the subgraph must be a clique.[3] Since a clique is also a subgraph, for given crosstalk constraints, the overhead of a memory-based code is no greater than that of a memoryless code.

We implemented the pruning algorithm in C and showed that the algorithm returns the codeword set for the optimal solution. Unfortunately, when the bus size is greater than 12-bits, the pruning requires a prohibitively large amount of memory and the computation time becomes too large. As an improvement, Victor in [99] proposed a class pruning algorithm specifically for the $3C$-free codes. The algorithm returns a sub-optimal solution. However, it allows the area-overhead be computed analytically. Figure 5.3 compares the overhead of $3C$-free FTF codes and memory-based codes produced through class pruning. It indicates that on average the memory-based code exhibits $\sim$15% less area overhead. For example. for the input bus size of 32, the

[3] A clique is a complete graph, in which every pair of vertices adjacent [95].

memory-based code requires only 40 wires, as opposed to 46 wires needed for a memoryless code.

5.3 Codeword Generation Using ROBDD

In this section, we present a method for memory-based code generation using Reduced Ordered Binary Decision Diagrams (ROBDDs) [22]. It is a two-step approach to determine the effective bus of width n that can be encoded in a kC-free manner, (for $1 \leq k \leq 4$) using a physical bus of width m:

- First, we construct an ROBDD $G_m^{kC-free}$ which encodes all vector transitions on the m-bit bus that are kC free.
- From $G_m^{kC-free}$, we find the effective bus width n, such that an n-bit data bus can be encoded in a kC free manner using $G_m^{kC-free}$.

5.3.1 *Efficient Construction of $G_m^{kC-free}$*

In the first step, we employ an ROBDD-based construction of $G_m^{kC-free}$. In particular, we inductively compute $\overline{G_m^{kC-free}}$. Since the ROBDD of a logic function and its complement contain the same number of nodes (except for a complement pointer), this enables an efficient construction of $G_m^{kC-free}$. Further, the ROBDD allows us to represent the legal (and illegal) kC free crosstalk transitions on the bus implicitly, share nodes maximally and in a canonical manner.

Suppose we want to construct $G_m^{kC-free}$. In this case, we allocate $2m$ ROBDD variables. The first m variables correspond to the vector from which a transition is made (referred to as $v = \{v_1, v_2, \cdots, v_m\}$). The next m variables correspond to the vector to which a transition is made (referred to as $w = \{w_1, w_2, \cdots, w_m\}$). If a vector sequence $v^* \rightarrow w^*$ is legal with respect to kC crosstalk, then $w^* \rightarrow v^*$ is also legal. In other words, $G_m^{kC-free}(v^*, w^*) = G_m^{kC-free}(w^*, v^*)$.

We construct the ROBDD for $\overline{G_m^{kC-free}}$ by using ROBDDs of intermediate, partially kC cross-talk free ROBDDs $\overline{G_i^{kC}}$ $(3 \leq i \leq m)$. The construction of the ROBDD of G_m^{kC} proceeds inductively, starting with the base case of G_3^{kC}. The construction of G_3^{4C}, G_3^{3C} and G_3^{2C} are described next.

$$G_3^{4C} = \left[\begin{array}{cccccc|cccccc} \multicolumn{6}{c|}{\overbrace{\hspace{5cm}}^{v}} & \multicolumn{6}{c}{\overbrace{\hspace{5cm}}^{w}} \\ v_1 & v_2 & v_3 & v_4 & \cdots & v_m & w_1 & w_2 & w_3 & w_4 & \cdots & w_m \\ 1 & 0 & 1 & - & \cdots & - & 0 & 1 & 0 & - & \cdots & - \\ 0 & 1 & 0 & - & \cdots & - & 1 & 0 & 1 & - & \cdots & - \end{array} \right]$$

We initialize G_3^{3C} with the transitions of G_3^{4C}, and then append additional $3 \cdot C$ transitions:

$$G_3^{3C} = G_3^{4C}$$

$$G_3^{3C} + = \left[\begin{array}{cccccc|cccccc}
v_1 & v_2 & v_3 & v_4 & \cdots & v_m & w_1 & w_2 & w_3 & w_4 & \cdots & w_m \\
0 & 0 & 1 & - & \cdots & - & 0 & 1 & 0 & - & \cdots & - \\
0 & 1 & 0 & - & \cdots & - & 0 & 0 & 1 & - & \cdots & - \\
1 & 0 & 1 & - & \cdots & - & 1 & 1 & 0 & - & \cdots & - \\
1 & 1 & 0 & - & \cdots & - & 1 & 0 & 1 & - & \cdots & - \\
0 & 1 & 0 & - & \cdots & - & 1 & 0 & 0 & - & \cdots & - \\
1 & 0 & 0 & - & \cdots & - & 0 & 1 & 0 & - & \cdots & - \\
0 & 1 & 1 & - & \cdots & - & 1 & 0 & 1 & - & \cdots & - \\
1 & 0 & 1 & - & \cdots & - & 0 & 1 & 1 & - & \cdots & -
\end{array}\right]$$

(columns $v_1 \ldots v_m$ grouped under $\overbrace{\qquad}^{v}$, columns $w_1 \ldots w_m$ grouped under $\overbrace{\qquad}^{w}$)

Similarly, we initialize G_3^{2C} with the transitions of G_3^{3C}, and then append additional $2C$ transitions:

$$G_3^{2C} = G_3^{3C}$$

$$G_3^{2C} + = \left[\begin{array}{cccccc|cccccc}
v_1 & v_2 & v_3 & v_4 & \cdots & v_m & w_1 & w_2 & w_3 & w_4 & \cdots & w_m \\
0 & 0 & 1 & - & \cdots & - & 1 & 1 & 0 & - & \cdots & - \\
1 & 1 & 0 & - & \cdots & - & 0 & 0 & 1 & - & \cdots & - \\
1 & 0 & 0 & - & \cdots & - & 0 & 1 & 1 & - & \cdots & - \\
0 & 1 & 1 & - & \cdots & - & 1 & 0 & 0 & - & \cdots & - \\
0 & 1 & 0 & - & \cdots & - & 1 & 1 & 1 & - & \cdots & - \\
0 & 0 & 0 & - & \cdots & - & 1 & 0 & 1 & - & \cdots & - \\
1 & 1 & 1 & - & \cdots & - & 0 & 1 & 0 & - & \cdots & - \\
1 & 0 & 1 & - & \cdots & - & 0 & 0 & 0 & - & \cdots & - \\
0 & 0 & 1 & - & \cdots & - & 0 & 1 & 1 & - & \cdots & - \\
0 & 1 & 1 & - & \cdots & - & 0 & 0 & 1 & - & \cdots & - \\
1 & 0 & 0 & - & \cdots & - & 1 & 1 & 0 & - & \cdots & - \\
1 & 1 & 0 & - & \cdots & - & 1 & 0 & 0 & - & \cdots & - \\
0 & 1 & 0 & - & \cdots & - & 0 & 0 & 0 & - & \cdots & - \\
0 & 0 & 0 & - & \cdots & - & 0 & 1 & 0 & - & \cdots & - \\
1 & 0 & 1 & - & \cdots & - & 1 & 1 & 1 & - & \cdots & - \\
1 & 1 & 1 & - & \cdots & - & 1 & 0 & 1 & - & \cdots & - \\
0 & 0 & 1 & - & \cdots & - & 1 & 0 & 0 & - & \cdots & - \\
1 & 0 & 0 & - & \cdots & - & 0 & 0 & 1 & - & \cdots & - \\
0 & 1 & 1 & - & \cdots & - & 1 & 1 & 0 & - & \cdots & - \\
1 & 1 & 0 & - & \cdots & - & 0 & 1 & 1 & - & \cdots & -
\end{array}\right]$$

(columns $v_1 \ldots v_m$ grouped under $\overbrace{\qquad}^{v}$, columns $w_1 \ldots w_m$ grouped under $\overbrace{\qquad}^{w}$)

Note that the ROBDDs for $\overline{G_3^{4C}}$, $\overline{G_3^{3C}}$ and $\overline{G_3^{2C}}$ are *only partially free of* $4C$, $3C$ and $2C$ transitions. They are immune to $4C$, $3C$ and $2C$ transitions only on the first three bits of the bus. We utilize the complements of these ROBDDs to construct $\overline{G_m^{kC-free}}$ (for $k = 2, 3$ and 4). The ROBDD $G_m^{kC-free}$ is free of kC transitions for all m-bits of the bus.

Algorithm 5.2 describes the inductive construction of $\overline{G_m^{kC-free}}$ from G_3^{kC}. In this algorithm, $G_3^{kC}((v_i, v_{i+1}, v_{i+3}) \leftarrow (v_1, v_2, v_3), (w_i, w_{i+1}, w_{i+3}) \leftarrow (w_1, w_2, w_3))$ refers to the ROBDD variable substitution of (v_i, v_{i+1}, v_{i+3}) and (w_i, w_{i+1}, w_{i+3}) by (v_1, v_2, v_3) and (w_1, w_2, w_3) respectively.

Algorithm 5.2 Inductive Construction of $\overline{G_m^{kC-free}}$ from G_3^{kC}

for $i = 1$ *to* $m - 3$ **do**
  $G_{i+3}^{kC} = G_{i+2}^{kC} + G_3^{kC}((v_{i+1}, v_{i+2}, v_{i+3}) \leftarrow (v_1, v_2, v_3), (w_{i+1}, w_{i+2}, w_{i+3})$
  $\leftarrow (w_1, w_2, w_3))$
end for
$\overline{G_m^{kC-free}} \leftarrow G_m^{kC}$
return $\overline{G_m^{kC-free}}$

Note that only the final $\overline{G_m^{kC-free}}$ that is constructed using Algorithm 5.2 is utilized for CODEC construction, since the intermediate ROBDDs for G_i^{kC} $(i < m)$ will possibly have kC or greater crosstalk transitions, since they are expressed over $2m$ variables.

5.3.2 *An Example*

As an example, let us construct a $\overline{G_5^{4C-free}}$. We start by constructing G_3^{4C} as discussed earlier:

$$G_3^{4C} = \begin{bmatrix} \overbrace{\begin{matrix} v_1 & v_2 & v_3 & v_4 & v_5 \end{matrix}}^{v} & \overbrace{\begin{matrix} w_1 & w_2 & w_3 & w_4 & w_5 \end{matrix}}^{w} \\ \begin{matrix} 1 & 0 & 1 & - & - \end{matrix} & \begin{matrix} 0 & 1 & 0 & - & - \end{matrix} \\ \begin{matrix} 0 & 1 & 0 & - & - \end{matrix} & \begin{matrix} 1 & 0 & 1 & - & - \end{matrix} \end{bmatrix}$$

Next we run Algorithm 3.1. Two iterations of the **for** loop are required, and their intermediate results are:

$$G_4^{4C} = \begin{bmatrix} \overbrace{\begin{matrix} v_1 & v_2 & v_3 & v_4 & v_5 \end{matrix}}^{v} & \overbrace{\begin{matrix} w_1 & w_2 & w_3 & w_4 & w_5 \end{matrix}}^{w} \\ \begin{matrix} 1 & 0 & 1 & - & - \end{matrix} & \begin{matrix} 0 & 1 & 0 & - & - \end{matrix} \\ \begin{matrix} 0 & 1 & 0 & - & - \end{matrix} & \begin{matrix} 1 & 0 & 1 & - & - \end{matrix} \\ \begin{matrix} - & 1 & 0 & 1 & - \end{matrix} & \begin{matrix} - & 0 & 1 & 0 & - \end{matrix} \\ \begin{matrix} - & 0 & 1 & 0 & - \end{matrix} & \begin{matrix} - & 1 & 0 & 1 & - \end{matrix} \end{bmatrix}$$

and

$$G_5^{4C} = \left[\begin{array}{ccccc ccccc} \multicolumn{5}{c}{\overbrace{\hspace{10em}}^{v}} & \multicolumn{5}{c}{\overbrace{\hspace{10em}}^{w}} \\ v_1 & v_2 & v_3 & v_4 & v_5 & w_1 & w_2 & w_3 & w_4 & w_5 \\ 1 & 0 & 1 & - & - & 0 & 1 & 0 & - & - \\ 0 & 1 & 0 & - & - & 1 & 0 & 1 & - & - \\ - & 1 & 0 & 1 & - & - & 0 & 1 & 0 & - \\ - & 0 & 1 & 0 & - & - & 1 & 0 & 1 & - \\ - & - & 1 & 0 & 1 & - & - & 0 & 1 & 0 \\ - & - & 0 & 1 & 0 & - & - & 1 & 0 & 1 \end{array} \right]$$

The ROBDD of the above cover is returned by the routine as $\overline{G_5^{4C-free}}$.

5.3.3 Finding the Effective kC Free Bus Width from $G_m^{kC-free}$

If an n-bit ($n < m$) bus can be encoded using the legal transitions in $G_m^{kC-free}$, then there must exist a closed set of vertices $V_c \subseteq B^m$ in the v space of $G_m^{kC-free}(v, w)$ such that:

- Each source vertex $v_s \in V_c$ has at least 2^m outgoing edges (v_s, w_d) to destination vertices w_d (including the self edge), such that the destination vertex $w_d \in V_c$.
- The cardinality of V_c is at least 2^n.

The resulting encoder is memory-based.

Given $G_m^{kC-free}$, we find n using Algorithm 5.3. The input to the algorithm is n and $G_m^{kC-free}$. We first find the out-degrees (self-edges are counted) of each $v_s \in B^m$ in the v space of $G_m^{kC-free}$. This is done by logically ANDing the ROBDD of the vertex v_s with $G_m^{kC-free}$. We find the cardinality of the resulting ROBDD – it represents the out-degree of v_s. If the number of out-edges of any v_s is greater than 2^n, we add v_s (and its out-degree) into a hash table V.

For each $v_s \in V$, we next check if each of its destination nodes w_d are in V. If $w_d \notin V$, we decrement the out-degree of v_s by 1. If the out-degree of v_s becomes less than 2^m, we remove v_s from V.

These operations are performed until convergence. If at this point, the number of surviving vertices in V is 2^n or more, then an n-bit memoryless CODEC can be constructed from $G_m^{kC-free}$.

We initially call the algorithm with $n = m - 1$ (where m is the physical bus size). If an n-bit bus cannot be encoded using $G_m^{kC-free}$, then we decrement n. We repeat this until we find a value of n such that the n-bit bus can be encoded by $G_m^{kC-free}$.

Algorithm 5.3 Testing if $G_m^{kC-free}$ can encode an n-bit bus

test_encoder(n, $G_n^{kC-free}$)
find out $-$ *degree*(v_s) *of each node* v_s, insert (v_s, *out* $-$ *degree*(v_s)) in V if *out* $-$ *degree*(v_s) $\geq 2^n$
degrees_changed $= 1$
while *degrees_changed* **do**
  degrees_changed $= 0$
  for *each* $v_s \in V$ **do**
    for *each* w_d S.T. $G_m^{kC-free}(v_s, w_d) = 1$ **do**
      if $w_d \notin V$ **then**
        decrement out $-$ *degree*(v_s) in V
        degrees_changed $= 1$
      end if
      if *out* $-$ *degree*(v_s) $< 2^n$ **then**
        $V \leftarrow V \setminus v_s$
        break
      end if
    end for
  end for
end while
if $|V| \geq 2^n$ **then**
  print(n-bit bus may be encoded using $G_m^{kC-free}$)
else
  print(n-bit bus cannot be encoded using $G_m^{kC-free}$)
end if

5.3.4 Experimental Results

We ran our algorithm to construct $G_m^{kC-free}$ and to find the effective bus width n for $4C$, $3C$ and $2C$ free transitions. The ROBDD of $G_m^{kC-free}$ is constructed as described in Sect. 5.3.1, and the test for the effective bus width is performed as described in Sect. 5.3.3. Based on the values of m and the corresponding effective bus width n, we plot the overhead in Fig. 5.4.

Note that this figure indicates that the asymptotic bus size overheads for the memory based codes are much lower than the memoryless code overheads. The overhead for a $2C$ free memory based code is about 117% compared to 146% for a memoryless code. The overhead for a $3C$ free memory based code is about 30% compared to 44% for a memoryless code. The overhead for a $4C$ free memory based code is about 8% compared to 33% for a memoryless code.

The memory-based CODEC designs are more complicated. Bus partitioning can also be used in the memory-based CODEC designs, where a wider bus, is partitioned into small groups and encoded and decoded independently.

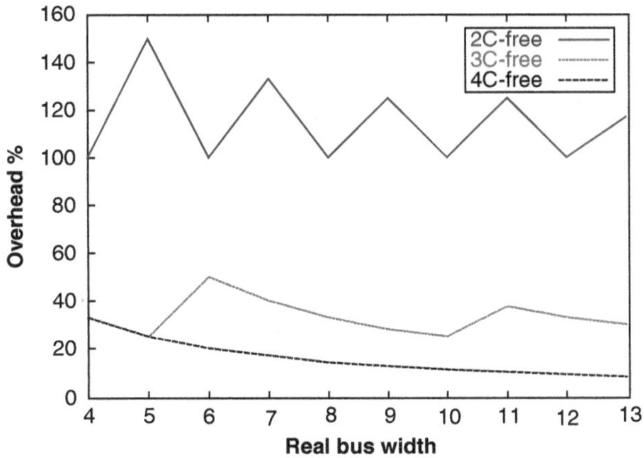

Fig. 5.4 Bus size overheads for Memory-based CODECs

The standard-cell based implementation of the encoder and decoder involves a few levels of logic, and its delay is estimated to be 250 ps. This delay can be reduced further by using Programmable Logic Arrays (PLAs) to implement the CODEC circuitry.

5.4 Summary

Memory-based CACs have been discussed in this chapter. Compared to memoryless codes, memory-based CACs have lower area overhead and therefore can be attractive. The biggest disadvantage of the memory-based code is the high complexity of their CODECs.

In Sect. 5.1, we presented a very simple 4C-free code design. It requires 33% overhead and the CODEC is very simple to implement. In Sects. 5.2 and 5.3, we discussed two general design techniques for memory-based code design: code pruning and the ROBDD-based approach.

In the code pruning approach, we first represent the bus transitions using a graph. Code pruning is equivalent to finding a subgraph that maximizes the minimal degree. Even though code pruning is used for memory-based code design, it can also be used for memoryless code design by constraining the resulting subgraph to a clique.

Codeword pruning returns codewords with optimal overhead. However, the high memory requirement and large run time of this approach make it impractical for large busses. Our experimental results show that memory-based 3C-free codes on average have a 15% lower overhead than their memoryless counterparts.

A second codeword searching algorithm we proposed is based on ROBDDs. In this approach, instead of explicitly enumerating all legal bus transitions, we enumerate only the illegal transitions inductively, and then derive the ROBDD for the legal transitions. The ROBDD-based approach represents the transitions in a compact and canonical fashion. The algorithm to find the effective kC free bus bandwidth n of a bus with physical width m was also presented.

Both approaches demonstrated that theoretically, the overheads can be as low as 117%, 30% and 8% for memory-based CODECs that are free of $4C$, $3C$ and $2C$ transitions respectively, which is much lower than the area overheads for the corresponding memoryless codes, which are 146%, 44% and 33% respectively.

Chapter 6
Multi-Valued Logic Crosstalk Avoidance Codes

So far, the discussions on on-chip crosstalk avoidance have been limited to binary valued busses. This can be attributed to the fact that binary-valued busses are almost exclusively used for on-chip interconnects in today's digital circuits thanks to their seamless ability to interface with the logic circuit. Further, binary-valued busses result in simple driver and receiver designs, with low power consumption. However, multi-valued busses have their advantages over the binary busses. In this chapter, we shift the focus of the discussion to how to built crosstalk avoiding multi-valued busses. We will show that multi-valued busses can out perform binary-valued busses in speed, power and area-overhead.

A multi-valued bus is a bus in which each digit can represent more than two logic values, as opposed to two values in a binary bus. Physically, these logic values are represented by different voltage levels or sometimes current levels. For illustration purposes, we will assume the voltage mode implementation unless otherwise specified. Figure 6.1 shows the waveform of a 4-valued bus wire. To achieve maximum noise margin, the circuit is implemented such that the voltage levels representing each value are evenly separated between 0V and V_{dd}. Therefore, for N-valued logic, the voltage values are $\{0, \frac{1}{N-1}V_{dd}, \frac{2}{N-1}V_{dd} \cdots \frac{N-2}{N-1}V_{dd}, V_{dd}\}$ and the voltage step is $V_{step} = \frac{1}{N-1}V_{dd}$.

Compared to a parallel binary-valued bus, the main advantage of a multi-valued parallel bus is that each wire carries more information and therefore the *bit density* is higher. The bit density of an N-valued logic signal is computed as $\rho(N) = \log_2 N$. For the 4-valued logic bus, every wire represents 2 binary bits since $\rho(4) = 2$. For a ternary (3-valued) bus, $\rho(3) = 1.585$ and for a decimal (10-level) bus, $\rho(10) = 3.32$. A higher bit density greatly reduces the number of wires needed for representing values in a given range. For bus interconnects, this often means fewer wires are required.

A widely used high-speed interconnect design technique is to reduce the signal swing. It improves signal speed since the delay is proportional to the voltage swing, and also reduces the power consumption. The related idea of multi-valued busses has been studied for high throughput interconnect as it offers higher bit density than binary busses [18, 71, 98, 103]. The 4-level pulse amplitude modulation (PAM) is a popular multi-valued logic data transmission approach [18, 103]. It offers sufficient

C. Duan et al., *On and Off-Chip Crosstalk Avoidance in VLSI Design*,
DOI 10.1007/978-1-4419-0947-3_6,

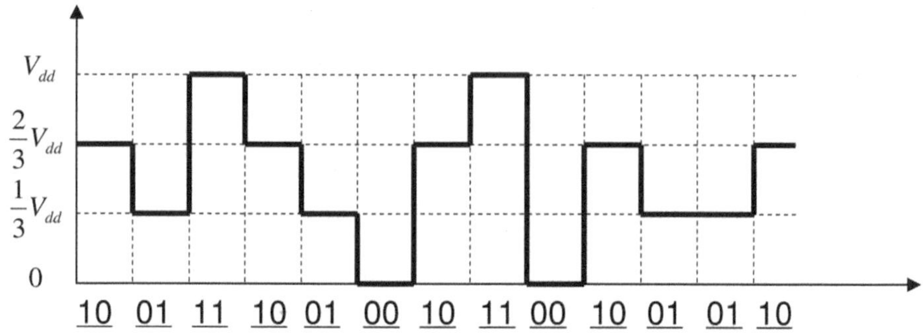

Fig. 6.1 Illustration of a 4-valued logic waveforms

noise margin and the number of logic levels is a power of two. However, crosstalk was not addressed in any of these papers and all these research has been strictly limited to off-chip interconnects.

6.1 Bus Classification in Multi-Valued Busses

In Chap. 2, we defined an *effective capacitance* for binary busses as

$$C_{eff,j} = C_L \cdot (1 + \lambda(2 - \delta_{j,j-1} - \delta_{j,j+1}))$$

and showed that both the energy consumption and signal delay of a data line can be expressed as functions of C_{eff}. For example, the delay on the jth line is given as

$$\tau_j = k \cdot abs(C_L \cdot \Delta V_j + C_I \cdot \Delta V_{j,j-1} + C_I \cdot \Delta V_{j,j+1}) \quad (6.1)$$

where ΔV_j is the voltage change on the jth line and $\Delta V_{j,k} = \Delta V_j - \Delta V_k$ is the relative voltage change between the jth and kth line. Assuming the output voltage levels are V_{dd} and 0V, we have $\Delta V_j \in \{0, \pm V_{dd}\}$ and $\Delta V_{j,k} \in \{0, \pm V_{dd}, \pm 2V_{dd}\}$.

If we let V_{step} be the voltage step between logic levels ($V_{step} = V_{dd}$ for a binary bus). Equation 6.1 can be rewritten as

$$\tau_j = k \cdot C_L \cdot V_{step} \cdot abs(\delta_j + 2\lambda \cdot \delta_j - \lambda \cdot \delta_{j,j-1} - \lambda \cdot \delta_{j,j+1}). \quad (6.2)$$

here $\delta_j = abs(\frac{\Delta V_j}{V_{step}})$ is the normalized signal transition and $\delta_{j,k} = \frac{sign(V_j) \cdot \Delta V_k}{V_{step}}$ is the normalized relative transition. For an N-level bus, $\delta_j = \{0, 1, \cdots, N-1\}$ and $\delta_{j,k} = \{0, \pm 1, \cdots \pm (N-1)\}$. When $N = 1$, the bus is binary, and $\delta_j = 1$ when there is a transition on the jth line, otherwise $\delta_j = 0$; $\delta_{j,k} = 1$ when both jth and kth lines transition in the same direction, -1 when they transition in opposite directions and 0 otherwise.

We define a *normalized total crosstalk* of the jth line ($X_{eff,j}$) as

$$X_{eff,j} = abs(2\delta_j - \delta_{j,j-1} - \delta_{j,j+1}), \quad (6.3)$$

and for a N-valued bus, we know

$$\min\{X_{eff,j}\} = 0$$
$$\max\{X_{eff,j}\} = 4 \cdot (N-1).$$

The maximum $X_{eff,j}$ occurs when $\delta_j = -\delta_{j,j-1} = -\delta_{j,j+1} = \pm(N-1)$, i.e., the wire in the middle transitions from one extreme value to the other extreme value and the two adjacent wires transition from one extreme level to the other extreme level in the opposite direction. For a binary bus, $\max\{X_{eff,j}\} = 4$. The relation between $X_{eff,j}$ and $C_{eff,j}$ can be expressed as

$$C_{eff,j} = (1 + \lambda X_{eff,j})C_L \approx X_{eff,j} \cdot C_I.$$

$\min\{X_{eff,j}\} = 0$ occurs when $\delta_{j,j-1} = \delta_{j,j+1} = \delta_j$. $\max\{X_{eff,j}\} = 4$ occurs when $\delta_{j,j-1} = \delta_{j,j+1} = -\delta_j$.

The signal delay on a multi-valued bus can be derived following the method give in Chap. 2. It is a function of $X_{eff,j}$ expressed as follows:

$$\tau_j = k \cdot C_L \cdot V_{step} \cdot abs(\delta_j + \lambda \cdot X_{eff,j}) \tag{6.4}$$

When $\lambda \gg 1$, Eq. 6.4 can be approximated as

$$\tau_j = k \cdot C_L \cdot V_{step} \cdot abs(\lambda \cdot X_{eff,j}) \tag{6.5}$$

Again, the delay τ_j is approximately linearly proportional to $X_{eff,j}$ and the maximum bus speed is limited by the value of $\max\{X_{eff,j}\}$. Based on the value of $X_{eff,j}$, we define the vector sequences on a N-level bus as 0X, 1X, 2X, ... 4(N – 1)X sequences and the bus can be classified as a 0X, 1X ... 4(N – 1)X bus, determined by the values of $\max_j\{X_{eff,j}\}$. A kX bus satisfies $\max_j\{X_{eff,j}\} \le k$.

The energy consumption is also a function of $X_{eff,j}$ and is given as

$$\begin{aligned} E_j &= C_L \Delta V_j \cdot V_{dd} - \lambda C_L (2\Delta V_j - \Delta V_{j-1} - \Delta V_{j+1}) V_{dd} \\ &= C_L \cdot V_{step} \cdot V_{dd} \cdot abs(\delta_j + \lambda \cdot X_{eff,j}) \\ &= E_j^L + \lambda E_j^I \end{aligned} \tag{6.6}$$

Equation 6.6 again shows that the energy consumption for each bit is separated into two parts: E_j^L is the energy to charge/discharge the load capacitance, and E_j^I the energy to charge/discharge the capacitance to adjacent wires. Again, the first term is negligible for DSM processes. The total energy consumption of the bus is given as

$$\begin{aligned} E_{total} &= \sum_{j=1}^{n} E_j^L + \lambda \sum_{j=1}^{n} E_j^I \\ &= \sum_{j=1}^{n} (1 + X_{eff,j} \cdot \lambda) C_L \cdot \Delta V_j^2 \end{aligned} \tag{6.7}$$

6.2 Low Power and Crosstalk Avoiding Coding on a Ternary Bus

The above equations give the signal delay and power consumption for busses with an arbitrary number of levels, N. A large N, however, results in reduced noise margins. In today's technology, the supply voltages have scaled down to 1V or lower. To guarantee sufficient noise margins, N is generally limited to a small value in practice. In this section, we will discuss crosstalk avoidance in ternary busses. A voltage-mode ternary bus has three output levels: V_{dd}, $V_{dd}/2$ and 0V and therefore the signal swing $V_{step} = V_{dd}/2$. δ_j, $\delta_{j,j+1}$ and $\delta_{j,j-1}$ vary within $[-2, 2]$. The maximum value of X_{eff} can be as high as 8.

Notation: In the following discussion, the three logic values on the ternary bus are denoted in bold as **+1, 0, −1**, representing the high, middle and low values respectively. For clarity, they are sometimes simplified as **+, 0, −**. Binary values are marked with underlines. For example, *101* indicates a 3-bit binary value and **+0+** is a 3-bit ternary value. Also, '⇒' denotes a mapping operation and '→' indicates a transition.

6.2.1 Direct Binary-Ternary Mapping

As shown before, a ternary bus has a higher "bit density" if it is used to represent a true ternary vector, as each ternary bit can represent 1.585 binary bits. Hence, an n-bit ternary bus can be used to replace an m-bit binary bus if $n \geq \lceil 1/\log_2 3^n \rceil \approx 0.63\, m$. However, the conversion between binary and ternary logic is complex and expensive to implement in VLSI. Thus a true ternary bus is rarely used for interconnect. Instead, we propose a ternary bus interconnect for binary logic based on a direct bit-to-bit mapping scheme. The mapping between the binary logic and ternary-valued bus is as follows:

1. Each binary bit is mapped directly to a line on the ternary bus, hence $n = m$;
2. A binary 0 is always mapped to a middle value on the ternary bus. i.e., 0⇒ **0**;
3. A binary 1 is mapped to either high or low value on the ternary bus. i.e. 1⇒ + or 1⇒ −.

Such a direct mapping scheme offers two advantages: it makes the encoding/decoding logic simple, and it also offers flexibility on the polarity for the binary 1. For example, the ternary vectors **+0+**, **−0−**, **+0−** and **−0+** all represent the same binary value 101. Table 6.1 gives several examples of ternary transition sequences and the corresponding X_{eff} on the middle bit. Notice that in this table, **+0+→0+0** produces a 4X crosstalk (on the middle line). For **+0+→0−0**, $X_{eff} = 0$. In our mapping scheme, both these ternary sequences represent the same binary sequence 101→ 010. Similarly, the **000→0++** transition has much less crosstalk (1X) compared to **000→0+−**, which encodes the same binary sequence but produces 4X crosstalk. In the extreme case, +−+→+−+ is an 8X sequence but a +++→ −−− is a 0X sequence, and both represent the same binary sequence.

Table 6.1 Examples of total crosstalk in a ternary bus

$\mathbf{V}_{t-1}$	$\mathbf{V}_t$	X_{eff}
000	+++	0
000	**0**++	1
000	**0**+−	3
+**0**+	**0**+**0**	4
+**0**+	**0**−**0**	0
−+**0**	+−**0**	6
+−+	−+−	8
+++	−−−	0

If we define the *bit polarity* as the sign of the ternary representation for a binary '1̲', the above comments show that the bit polarity can affect crosstalk significantly. The following two coding schemes exploit the flexibility in our mapping scheme to reduce the crosstalk, thereby yielding energy savings and an increase in speed.

6.2.2 4X Ternary Code

We first show the construction of a ternary sequence that eliminates 5X and higher crosstalk. We call this a "4X ternary code" since it satisfies

$$\max\{X_{eff,j}\} \leq 4, \forall j \in [1, n]$$

Let $\underline{\mathrm{b}}_j$ be the jth bit of the input vector, P_j the polarity and D_j the magnitude of the corresponding ternary representation. The following are the rules for constructing the 4X ternary sequence:

1. On any bit, direct '+→ −' or '− → +' transitions are prohibited.
2. $\underline{1} \rightarrow \underline{0}$ is mapped as +→**0** or − →**0**.
3. For a $\underline{0} \rightarrow \underline{1}$ transition on $\underline{\mathrm{b}_j}$, if $\underline{\mathrm{b}_{j-1}}$ transitions, $\mathbf{P}_j$ is coded so both lines transition in the same direction.
4. For a $\underline{0} \rightarrow \underline{1}$ transition on b_j, if b_{j-1} does not transition and b_{j+1} transitions from 1 to 0, $\mathbf{P}_j$ is coded so that the jth and the $(j+1)$th lines transition in the same direction.
5. For a $\underline{0} \rightarrow \underline{1}$ transition on b_j, if no transition occurs on either neighbor, $\mathbf{P}_j$ is coded so $\{\mathbf{P}_j = \mathbf{P}_{j-1}$ or $\mathbf{P}_j = \mathbf{P}_{j+1}\}$ with $\mathbf{P}_j = \mathbf{P}_{j-1}$ having the higher priority.

The first rule guarantees a maximum of 4X crosstalk on any bit of the ternary bus. Compared to a ternary bus without crosstalk avoidance coding where $\max\{X_{eff,j}\} = 8$, our coding will boost the bus speed by approximately a factor of 2. The additional rules allowing the encoder to choose the bit polarity to minimize local crosstalk whenever it is possible. This results in total crosstalk reduction and therefore lowers the bus power consumption. Table 6.3 gives an example of the output sequence produced of the 4X ternary code for a given binary sequence.

Table 6.2 Ternary driver truth table

Magnitude D_j	Polarity P_j	Logic value t_j	Voltage
0	X	0	V_0
1	0	$-$	V_-
1	1	$+$	V_+

Table 6.3 4X ternary sequence example

Binary	Ternary	X_{eff}
11110111	++−000−+	
00110101	00−−0+0+	01100121
11100011	++−000−+	01220111
01010100	0+0+0+00	10112122
10101110	−0−0−+−0	00001021
01110001	0+−+000−	01212200
00000011	000000−−	13431121
00011110	000+++−0	00110121

A CODEC can be easily implemented to satisfy all the rules above. As illustrated in Fig. 6.2, the inputs of the encoder are the binary vector $\underline{b_1 b_2 \ldots b_n}$ and the output is a ternary vector $\mathbf{t}_1\mathbf{t}_2\ldots\mathbf{t}_n$. Each line driver circuit can be considered of having two parts, the polarity encoder and the ternary driver. The ternary driver outputs the 3-level signal $\mathbf{t}_j$ based on the polarity ($\mathbf{P}_j$) and magnitude ($\mathbf{D}_j$) using the truth table given in Table 6.2. The jth polarity encoder has inputs of $\underline{b_{j-1} b_j b_{j+1}}$ and $\mathbf{P}_{j-1}\mathbf{P}_j\mathbf{P}_{j+1}$. Since the encoders for all lines are identical except for the boundary bits $\underline{b_0}$ and $\underline{b_n}$, a very efficient implementations can be realized due to the regularity. Figure 6.2 illustrates such an encoder.

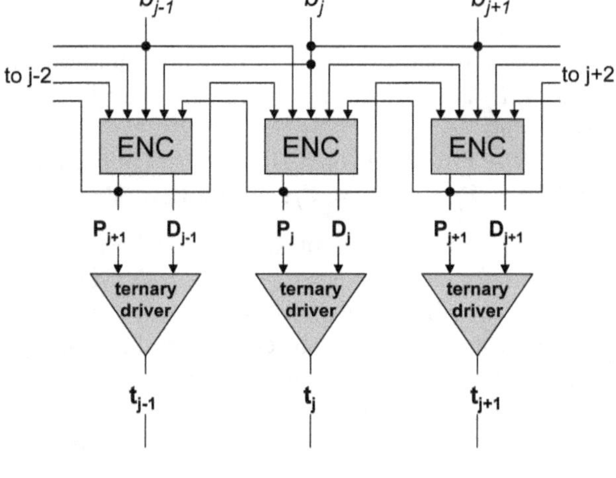

Fig. 6.2 4X encoder and driver circuit

6.2.3 3X Ternary Code

The 4X-code lowers the probability of 4X crosstalk significantly as we can see from Table 6.3. The maximum speed of a 4X-code bus is still limited by $X_{eff} = 4$. To further improve the bus speed, we propose a modified coding scheme that eliminates the 4X crosstalk completely. The coding scheme is called "3X ternary code" as it satisfies

$$\max\{X_{eff,j}\} \leq 3, \forall j \in [1, n]$$

The 3X code is constructed by partitioning the bus into 5-bit lanes (groups) and inserting a grounded wire between lanes as a shield. Now we apply the same 5 rules given in Sect. 6.2.2. Compared to the 4X code, such a design will have up to 20% area overhead. Since the 3X code can speed up the bus by ~33% compared to a 4X bus, this coding scheme offers an additional 11% gain in the throughput per unit area.

The proof that such codec cancels all 4X and higher crosstalk is given as follows:

Proof

1. The prohibition of a direct $+ \leftrightarrow -$ transition eliminates all the crosstalk over 4X since the maximum crosstalk on each side of the jth wire cannot exceed 2X.
2. With the $+ \leftrightarrow -$ transition prohibited, there are 16 transitions that can produce 4X crosstalk on the line in the middle. Each pair exhibits 2X crosstalk on either side. The full list is not given here to save space.
3. Due to the flexibility given to the polarity of a binary '1̲', we can remove all but the pairs with **XXX→X00** patterns. There are four such pairs: $+-+\rightarrow$**000**,$-+-\rightarrow$**000**, **0**$-+\rightarrow -$**00** and **0**$+-\rightarrow +$**00**. They can be generalized in two cases: **ABA→000** and **0AB→A00**, where **A** and **B** represent two values of opposite polarities.
4. In the case of **ABA→000**, we can prove that it is possible to prevent **ABA** from appearing on a bus that is no more than 5-bits wide. Considering a 3-bit slice $\mathbf{t}_j\mathbf{t}_{j+1}\mathbf{t}_{j+2}$ on the bus, since no direct transitions between **A** and **B** are allowed, the **ABA** pattern can only be formed from patterns that contain at least one **0**. Let the polarity of $\mathbf{t}_{j+1}$ and $\mathbf{t}_{j+2}$ be selected based solely on its left neighboring bits. If either $\mathbf{t}_{j+1}$ or $\mathbf{t}_{j+2}$ in the previous cycle has the value of **0**, this rule guarantees that an **ABA** will not be formed. Therefore the only other scenario we need to examine is the **0BA→XBA** transition. If $\mathbf{t}_j$ can be coded as either **A** or **B,** we can avoid **ABA**. This flexibility on $\mathbf{t}_j$ can be guaranteed if the X_{eff} between $\mathbf{t}_{j-2}$ and $\mathbf{t}_{j-1}$ is no more than 1 (so the total X_{eff} on $\mathbf{t}_{j-1}$ is no greater than 3). Such is the case if $\mathbf{t}_{j-2}$ is the left boundary bit. $\mathbf{t}_{j-2}\mathbf{t}_{j-1}\mathbf{t}_j\mathbf{t}_{j+1}\mathbf{t}_{j+2}$ form a 5-bit bus and $\mathbf{t}_{j-2}$ is the left boundary bit.
5. Similarly, **0AB→X00** can be coded as **0AB→B00** if the polarity of $\mathbf{t}_j$ is flexible. This requires that the X_{eff} between $\mathbf{t}_{j-1}$ and $\mathbf{b}_{j-2}$ is less than or equal to 1. Again, it can be guaranteed when $\mathbf{t}_{j-2}$ is the left boundary bit. □

The 3X codec is less regular than the 4X codec. In each 5-bit lane, the polarity encoder for the center bit has higher complexity than encoders for other bits. Each

5-bit lane is independent from others and therefore there is no ripple delay throughout the entire bus during encoding.

6.3 Circuit Implementations

Binary on-chip busses are almost exclusively implemented in voltage mode. The driver is an inverter that is appropriately sized to drive the load. The receiver is normally a minimum sized inverter with the switching point set at $V_{dd}/2$.

The ternary busses can be implemented in either voltage mode or current mode. A current mode bus implementation is illustrated in Fig. 6.3a. The driver consists of two current sources that are switched on/off to output $2I_d, I_d$ or 0. A low impedance termination is need at the receiver end. The receiver consists of two current comparators. The outputs from the current comparators are decoded by a simple decoder which consists of an OR gate and an inverter.

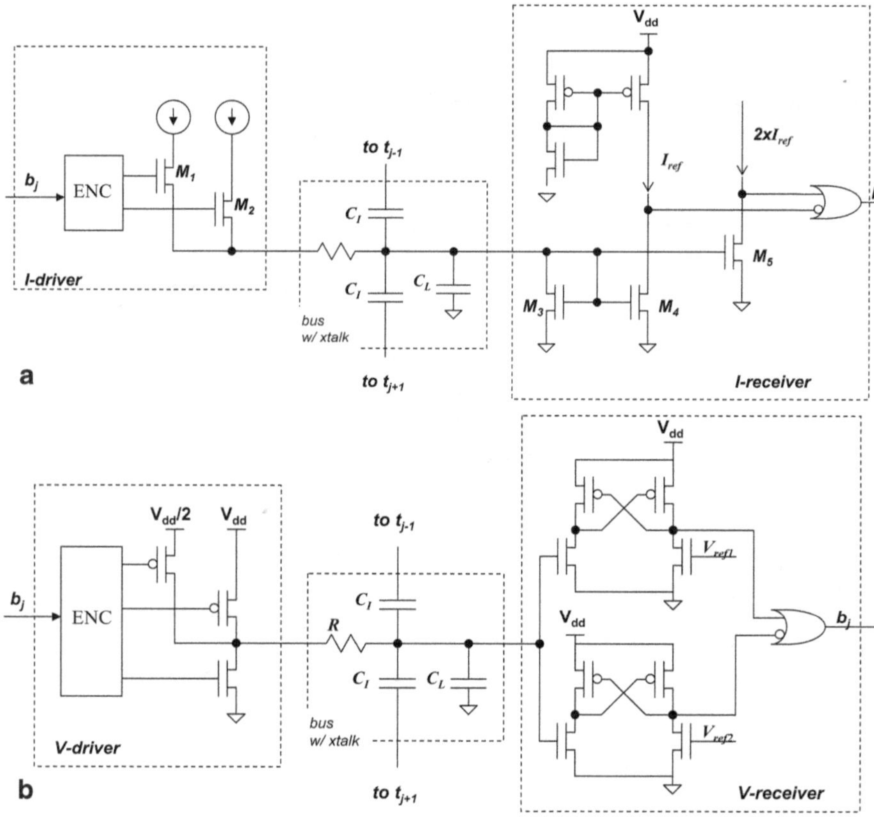

Fig. 6.3 Current-mode and Voltage-mode busses **a** Current-mode bus driver and receiver **b** Voltage-mode bus driver and receiver

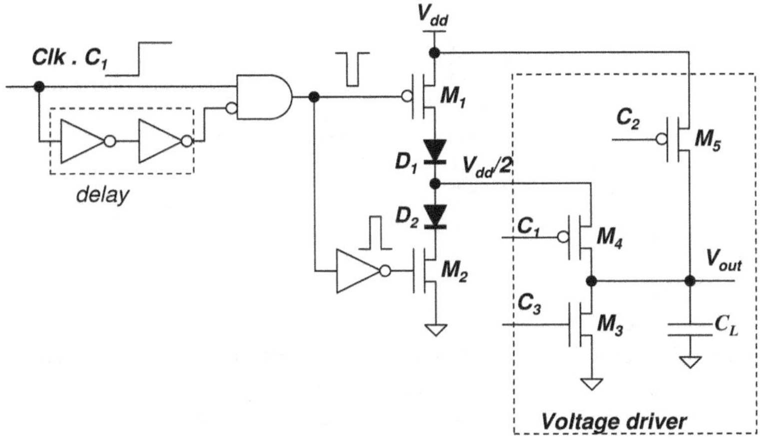

Fig. 6.4 On-chip $V_{dd}/2$ generation

The current mode drivers and receivers are less sensitive to supply voltage variations. However, the power consumption is generally higher than the voltage mode circuit since a static current is required to drive the low impedance load.

A voltage-mode bus, on the other hand, typically consumes less power. An implementation is shown in Fig. 6.3b. The voltage driver consists of two PMOS devices and one NMOS device. The three output voltages are set as: $V_+ = V_{dd}$, $V_0 = V_{dd}/2$ and $V_- = 0V$. The receiver consists of two voltage comparators and the outputs from the comparators are decoded by a simple decoder. Two voltage references $V_{ref1} = \frac{V_+ + V_0}{2} = \frac{3V_{dd}}{4}$ and $V_{ref2} = \frac{V_- + V_0}{2} = \frac{V_{dd}}{4}$ can be generated on-chip and shared by all the receivers.

The voltage-mode driver requires a low impedance supply voltage V_0 ($V_{dd}/2$ in our case). In the case a separate power supply is not available off-chip, it can be implemented on chip. Figure 6.4 shows a simple and yet efficient circuit that generates the $V_{dd}/2$ supply. When both M1 and M2 are turned on, two forward biased diodes D1 and D2 create a low-impedance $V_{dd}/2$ supply. To minimize the static current, M1 and M2 are turned on only for a brief period (when the transition occurs and a large current is drawn). Due to crosstalk, it is necessary to turn on M1 and M2 even when the input is steady high. The turn-on time τ is determined by the maximum rise/fall time τ on the bus. Such a short delay can begenerated using inverters. M4 is optional.

6.4 Experimental Results

We first verified the maximum crosstalk and total energy consumption of our ternary codes using randomly generated vectors on 5, 8, 16 and 32-bit busses. Our simulation confirmed that we can achieve a maximum crosstalk of 4X for 8, 16 and 32-bit busses and 3X crosstalk for the bus of 5-bit.

Table 6.4 Bus performance comparison

Bus type	4XT	3XT	SB	HB	RP	TT
EF (10^4)	6.13	6.67	19.7	8.38	12.1	7.55
MaxDelay	4X	3X	4X	4X	8X	8X
PDP(10^5)	2.45	2.00	7.88	3.35	9.68	6.04
Pwr saving (%)	68.9	66.1	0	57.5	38.6	61.7
PDP gain (%)	68.9	74.6	0	57.5	−22.8	23.4
Bus area	1	1	1.97	1	1	0.68

We first compare our 4X ternary bus with a half swing binary bus. Both busses have a maximum crosstalk of 4X. Our simulations shows that for bus sizes of 5-bit, 8-bit, 16-bit and 32-bit, the 4X ternary code offers energy saving of 34.52%, 28.21%, 27.25% and 27.56% respectively. The reason for this saving is that our coding also minimizes the occurrence of high crosstalk (3X and 4X) transitions. The codec circuit power consumption is less than 2% of the total power consumption for a 5 mm bus.

Table 6.4 compares the performance of the 4X and 3X ternary bus against some other busses.[1] All busses in the table are 16-bit wide except the true ternary bus (which is 11-bit wide). Also included in the table is a full-swing binary bus with passive shielding since it is widely used in crosstalk-free bus designs. Note that for such a bus, there are 31 wires for a 16-bit bus. Also included in the simulation is a random polarity ternary bus (RP bus), which follows the bit-to-bit mapping rules with polarity assigned randomly. The lengths of the busses used in the simulation were 5 mm. The total power consumption includes the power consumed by drives, receivers and CODEC circuit (if applicable), which are implemented using a 65 nm process [7]. The sizing of the ternary drivers is done so that the 0 to $0.25V_{dd}$ delay is equal to the 0 to $0.5V_{dd}$ delay for the full swing drivers. The bus areas in the table are all normalized to the area of a 5 mm, 16-wire bus. A total of 10,000 randomly generated vector sequences are used for all the simulations. **EF** is defined as *normalized total energy* which is the E_{total} given in Eq. 6.6 normalized by setting $C_I = 1$ and $V_{step} = 1$. As we can see, the 4X ternary bus has the minimum energy consumption. It saves on average 68% over a full-swing bus with passive shield and 27% over a half-swing binary bus. It is interesting to see that even though the true ternary bus has only 11 lines, it consumes ~19% more power than the 4X ternary bus. The 3X ternary code consumes about 8% more energy than the 4X ternary bus due to the coupling crosstalk to the added shielding lines. We use *Power-maxDelay Product* (PDP) as the figure of merit for measuring the overall bus performance. PDP is the product of normalized total energy and the normalized maximum delay. Table 6.4 shows that the 3X ternary code has the minimal PDP while the full-swing binary bus with passive shielding has the highest PDP.

[1] Bus types: 4XT- 4X ternary; 3XT- 3X ternary; SB- full swing binary bus with passive shielding; HB- half swing binary; RP- random polarity ternary; TT- true ternary.

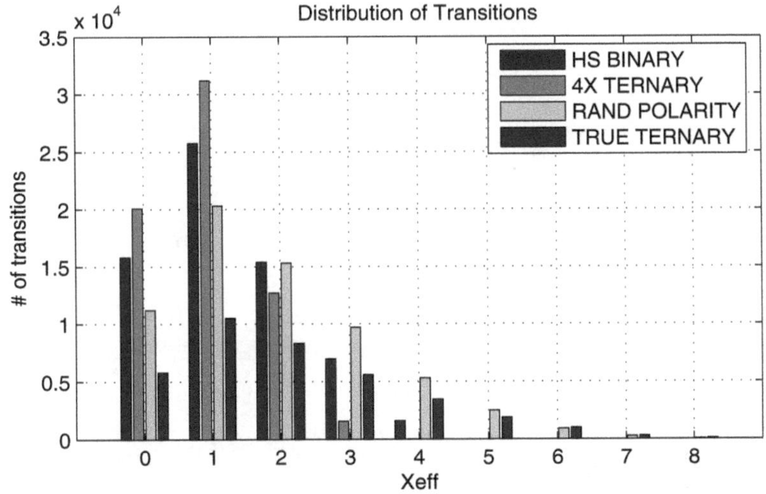

Fig. 6.5 Crosstalk distributions

Figure 6.5 shows the distribution of the crosstalk for several different types of busses. We can see that by applying our coding scheme, the distribution of the transitions shift towards lower X_{eff} values. The true ternary bus has fewer total number of transitions compared to the other busses since there are fewer bits on the true ternary bus (11-bits). However, a significant portion of the transitions have 4X and higher crosstalk. The total energy consumption is therefore higher than either coded ternary bus.

We implemented the voltage-mode ternary bus and verified the delay for different X_{eff} values in SPICE simulation. The results are given in Table 6.5 and as expected, we note that the delay scales roughly linearly with X_{eff}. Again, we use a 65 nm CMOS process model in the simulation [7], with wiring dimensions and parasitics obtained

Table 6.5 Delay vs. Crosstalk

Driver size	30X	30X	60X	60X
Bus length	5 mm	10 mm	10 mm	20 mm
Crosstalk	Delay (ns)			
0X	25	110	60	65
1X	110	220	140	200
2X	190	350	230	350
3X	300	580	350	590
4X	420	790	460	760
5X	520	970	530	960
6X	630	1140	640	1130
7X	730	1260	750	1290
8X	810	1390	850	1430

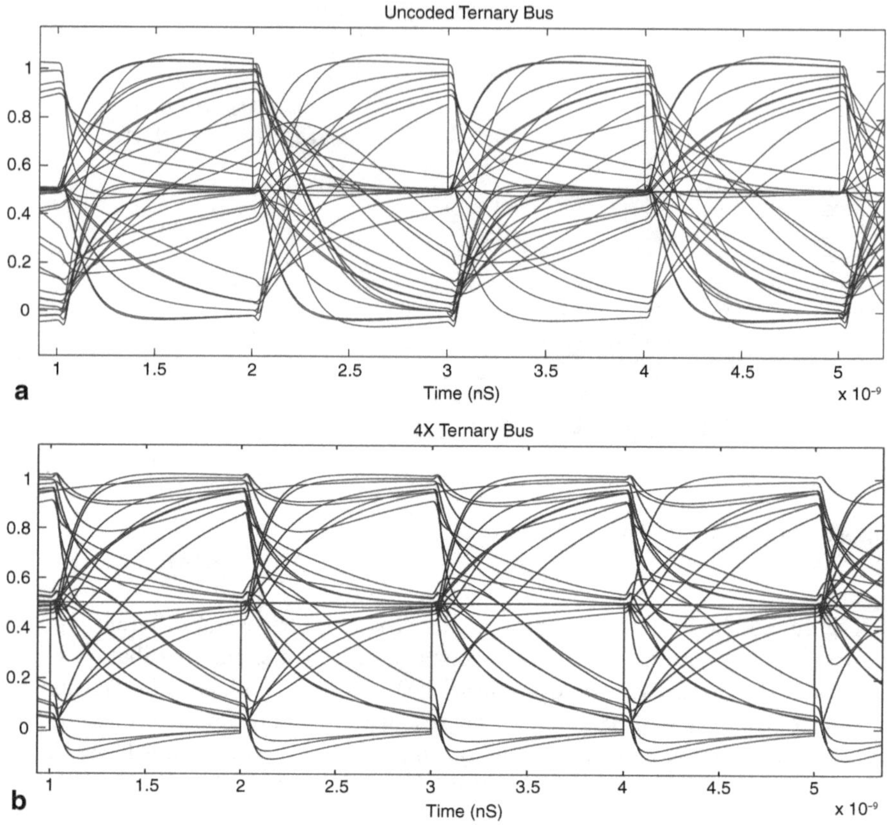

Fig. 6.6 Ternary bus eye-diagrams **a** Uncoded ternary bus **b** 4X ternary bus

from [61]. Figure 6.6 shows the eye diagrams for a uncoded and a 4X ternary bus. Note that the eye is more open for the 4X ternary bus.

6.5 Summary

At the beginning of this chapter, we derived delay and power consumption expressions for multi-valued busses and generalized the crosstalk-based bus classification for such busses. We then designed two ternary bus coding schemes to limit maximum crosstalk on a ternary bus. Both coding schemes simultaneously lower the power consumption and boost the speed of on-chip bus interconnects. The proposed 4X bus effectively improves the bus speed by ~100% compared to a true ternary bus or a full-swing binary bus, with no area overhead compared to a binary bus. The 3X bus offers an additional 33% speed increase with a 20% area overhead.

Our simulation results show that the 4X ternary bus with the proposed crosstalk avoidance coding schemes exhibits on average over 27% power saving over a binary bus with half swing, and over 68% lower power consumption than a full swing binary bus with passive shielding. Compared to a true ternary bus, the coded ternary bus can operate at over twice the speed and exhibits over 11% power saving. The encoder and decoder for both the proposed coding schemes can be implemented with very simple logic thanks to the simplicity and regularity of the coding logic and the use of direct bit-to-bit mapping between binary data words and ternary codes.

We also presented low power designs for the ternary driver and receiver circuits. We also proposed an efficient low impedance on-chip power supply generator, that when used with our voltage mode bus, reduces the static power consumption.

Chapter 7
Summary of On-Chip Crosstalk Avoidance

Conventional wisdom doubts that bus encoding would ever become a viable solution to the problem of inter-wire capacitive crosstalk due to its area overhead and CODEC complexity associated with such techniques.

In Part I, we presented some of the recent research in overcoming these obstacles and showed that such techniques can be highly efficient.

Following a brief overview of techniques in crosstalk avoidance at the beginning of Part I, we examined the phenomenons of crosstalk in on-chip busses and illustrated that capacitive coupling between adjacent parallel bus wires is the main contributor to the speed and power degradation in deep sub-micron signal busses. We introduced the notion of *effective capacitance* and showed that both delay and energy consumption of busses are linearly proportional to it. We introduced a bus classification system for binary busses to make the subsequent discussion clear. This classification system was later generalized to multi-valued busses.

With the establishment of this classification system, we introduced our first set of crosstalk avoidance codes—memoryless codes. These codes have the inherent advantage of requiring simpler CODECs. We showed that there exists a trade off between the area overhead and the speed-up factor for all the codes. Generally, the higher the speed-up, the larger the area overhead. Some of the memoryless codes we discussed include the FPF code (3C-free) [36], FTF code (3C-free code proposed by B. Victor [100]), 2C-free code and 1C-free configurations [35]. We showed that mathematical induction based approaches can be used to effectively generate codewords for both 3C-free codes and the 2C-free code. Such inductive approaches also allow us to compute the code cardinality without explicitly enumerating all the codewords. We found that based on our computation, 3C-free memoryless codes require $\sim$44% overhead, which is significantly lower than the conventional static shielding approaches. The overhead is higher for 2C-free codes but still lower than shielding techniques.

As part of a complete solution for memoryless codes, we also discussed in detail some efficient CODEC design approaches for these codes. The bus-partition based solution is simple and allows us to control the CODEC size and speed. Our solutions based on the Fibonacci Numeral System are advantageous over conventional solutions because of their significantly reduced area, simple design,

C. Duan et al., *On and Off-Chip Crosstalk Avoidance in VLSI Design*,
DOI 10.1007/978-1-4419-0947-3_7,

deterministic mapping, reusable modular structure and most importantly, the extreme low complexity and high speed.

We also investigated memory-based codes and showed that they are more efficient in terms of area overhead. A 4C-free code design was given as one of the simplest memory-based code, with a 33% overhead. We showed that graph pruning can be used to represent codewords and transitions. Based on a graph representation, we showed that codeword pruning is a simple way of finding the minimum encoding overhead. It, however, requires a large amount of memory and run time. We proposed an ROBDD-based approach that represents all the illegal code words canonically and generates codewords efficiently. It also suffers from the same limitations as the codeword pruning technique. As a result, realistic memory based CACs would need to partition large busses into smaller groups.

Finally, we investigated the crosstalk avoidance coding for multi-valued busses. We first generalized the classification system into the multi-valued realm and showed that the speed and power of a bus are related to the effective crosstalk coefficient, X_{eff}, a parameter defined in a manner similar to the effective capacitance C_{eff} in the binary bus case. Noting that the supply voltages are scaling down continuously, using more than 4 values in a multi-valued bus is impractical. Hence, we proposed two code designs specifically for the ternary bus. These codes use a bit-to-bit direct mapping which makes the encoding and decoding trivial. By allowing flexibility on the signal polarity and establishing a few coding rules, the encoded busses eliminate bus patterns with high crosstalk and significantly reduce the probability of other high crosstalk patterns. These codes exploit the low signal swing of the multi-valued signals. As a result, we were able to achieve a significant speed improvement and power saving.

We believe that the issue of crosstalk avoidance in on-chip interconnect to reduce delay and power will attract more and more attention as the technology marches ahead. Beyond the research presented in this monograph, there is room for further improvement. For example, efficient CODEC design for memory-based codes is a subject worth further investigation. Multi-valued logic based bus interconnect is another area that holds significant promise. Crosstalk in serial busses is also an important topic. Even though there have been several publications on combining crosstalk avoidance codes with other codes such as forward error correction codes, an practical, efficient design has yet to be discovered.

Some of the techniques discussed are not limited to bus crosstalk avoidance, instead, they can be applied to other areas. For example, the FPF code can be of particular interest to certain applications in digital communication, where certain patterns are designated to carry higher priority messages, such as emergency massages. Our FPF codes have 44% overhead, which is much less compared to existing schemes. The FPF code may also be used in serial data communications, in which back to back toggles are not allowed. Another potential applications is interference management in wireless communications. The transmission in a cell (or sector) is considered as interference to an adjacent cell (or sector). The accumulated interference can greatly reduce the capacity of the victim cell. It is conceivable that some of the crosstalk avoidance schemes may be applied for interference management. Another interesting property of the FPF code lies in its spectrum. Since the high frequency component are suppressed, we expect that by applying FPF encoding, high bandwidth efficiency can be achieved. This is extremely valuable for many wireless communication systems.

Part II
Off-Chip Crosstalk and Avoidance

Chapter 8
Introduction to Off-Chip Crosstalk

Integrated circuit (IC) performance has increased at an exponential rate since the first patent was issued to Robert Noyce of Fairchild Semiconductor in 1961 [10]. Since the advent of the IC, the number of transistors on an integrated circuit has roughly doubled every 18 months. This trend, also known as Moore's Law [96], has been consistently met over the past 45 years. This increase in system performance on the IC is predicted by the International Technology Roadmap for Semiconductors (ITRS) [11] to continue to follow Moore's Law into the foreseeable future. Historically, the gate delay of the digital circuitry has limited IC performance [59]; however, over the past decade the bottleneck of system performance has shifted from the gate delay of the integrated circuit to the package parasitics [96]. While package performance has steadily improved, it has not kept pace with the increases in integrated circuit performance. Package performance is predicted by the ITRS to only double over the next decade. This imbalance in performance expectations between the IC and the package is a major concern for system designers and for the continuation of Moore's Law.

The limitation in package performance comes from the parasitic inductance and capacitance in the electrical interconnect [32, 57, 96]. Package interconnect has historically been designed to meet mechanical, thermal, and cost objectives. In the past, the electrical performance of the interconnect was not an issue since it caused a relatively small performance degradation relative to the gate delay of the IC transistors [59]; however, with the dramatic improvement in transistor gate delay, package performance is now the leading determinant of the overall system performance.

8.1 The Role of IC Packaging

The purpose of an IC package is to electrically and mechanically connect the integrated circuit substrate to the system substrate. The most common substrate used in Very Large Scale Integrated (VLSI) circuitry is silicon [59]. Silicon offers a host of electrical advantages that allow the implementation of large numbers of transistors in a cost-effective and easy-to-manufacture manner. In VLSI silicon designs, conductors are most commonly implemented using Polysilicon ($p-Si$), Aluminum (Al), and Copper (Cu). Insulating layers in VLSI designs are typically implemented using

C. Duan et al., *On and Off-Chip Crosstalk Avoidance in VLSI Design*,
DOI 10.1007/978-1-4419-0947-3_8,

Silicon Oxide (SiO_2) and Silicon Nitride (Si_3N_4). The photolithography and deposition processes used in IC fabrication allow extremely small feature sizes to be printed on the substrate. Currently, feature sizes as small as 65 nm are being successfully implemented using VLSI processes [84].

System level interconnect in digital systems is typically constructed using printed circuit board (PCB) technology [32]. PCBs typically use Copper as their conducting layer. Insulating layers are implemented using dielectric materials such as FR4, Nelco-13, *GETEK*®, and Teflon. PCBs are constructed using a lamination process. Modern lamination processes are capable of producing minimum feature sizes between 4 and 100 μm.

An IC package serves many purposes. The first is to electrically connect the leads on the IC to the corresponding leads on the system PCB. Since the feature sizes on the IC are much smaller than the feature sizes on the PCB, the IC cannot be mounted directly to the system PCB. The package serves as a *density translator* for the electrical signals that will connect extremely fine-pitch signals on the IC to coarser-pitched signals on the PCB.

The second purpose of the package is to protect the substrate of the IC. The IC substrate consists of a very thin crystal of silicon that has transistors and interconnect deposited on it. This substrate is brittle and can be easily fractured, leading to circuit failure [96]. The package serves as a mechanical barrier and hermetic seal between the silicon substrate and the outside environment. In this manner the substrate is isolated from contamination and humidity, which can also lead to IC failures. In addition, the package absorbs thermal expansion mismatches between the silicon substrate and the system PCB. If the thermal mismatches were completely absorbed by the silicon substrate, it would lead to stress fractures and circuit failure.

The third major purpose of IC packaging is to remove heat from the IC substrate. Modern ICs consist of millions of transistors that each consume current and dissipate power. The cumulative power dissipation of the transistors lead to the generation of extreme amounts of heat in a relatively small area. This high thermal density adversely effects circuit performance by increasing the gate delay and shortening the life of the devices on the IC [59]. The typical range of operating junction temperatures for modern VLSI designs is between 80°C and 120°C on the silicon substrate [40, 59]; however, modern microprocessors are projected by the ITRS to surpass the 100 Watt power dissipation mark within the next couple of years [40]. All this power is dissipated by substrates that range from 5 to 20 mm^2 in size [54, 62]. The package serves as a heat transfer system that removes the heat from the IC substrate. The heat is delivered to a package surface from which it can ultimately be absorbed by ambient air. By doing this, the package can keep the circuitry on the IC within the acceptable range of junction temperatures. This leads to consistent and predictable performance of the devices on the IC substrate.

Historically, the thermal and mechanical aspects of the package were the main focus of the package design [96]. Since the gate delay of the devices on the IC were the largest source of performance limitation, the electrical interconnect in the package was optimized for mechanical and thermal purposes. This approach has been successful for the past 40 years. Only recently has the electrical interconnect in the package become the bottleneck to system performance. This has occurred due to the dramatic advances in integrated circuit technology that has reduced the gate delay of the on-chip devices to the point where they no longer limit system performance. This shift as resulted in the package (which was originally optimized for mechanical and thermal rather than electrical consideration) becoming the major limiting factor in system level electrical performance.

The electrical interconnect limits performance due to the parasitic resistance, inductance and capacitance that is present in the interconnect structure [32, 57, 58]. Since the interconnect structures were originally created for mechanical properties, their parasitic inductance and capacitance can be considerable relative to the datarates that are currently being transmitted through the package. Package design is a slow and expensive process that has not been able to keep pace with the performance increase in core IC technology. This makes the task of altering the interconnect structures within the package (to increase electrical performance) very difficult. The majority of VLSI designs today cannot afford the cost and time associated with re-engineering a package [40]. This means that in most cases designers must live with the performance limitations in the package. Any technique that can increase system performance without changing the package design is of tremendous value to the VLSI community.

8.2 Noise Sources in Packaging

Electrically the package has two objectives. The first is to deliver power to the devices on the integrated circuit. The power resides on the system PCB and must be conducted through the package to the devices on-chip. The second objective is to connect signals on the IC to the signal paths on the system PCB. In both of these cases electrical current and voltage are carried using interconnect structures within the package that were not originally optimized for electrical performance. The parasitic resistance, inductance and capacitance of these structures limits how efficiently the power can be delivered and how fast the signals can be transmitted.

8.2.1 Inductive Supply Bounce

When inductance is present in the path that carries power to the devices on the IC, a voltage will form across this inductance based on the relationship:

$$V_L = L \cdot \left(\frac{di}{dt}\right) \tag{8.1}$$

In the case of power delivery, the inductance in Eq. 8.1 is the parasitic inductance in the interconnect structure that carries current to and from the devices on-chip. $(\frac{di}{dt})$ is the rate of change of the current that is carried through the interconnect. As more and more transistors are integrated on-chip, the total amount of current that is being carried through the interconnect increases. In addition, as the operating frequency of the digital circuits increase, the time in which this current is delivered is reduced. This results in significant increases in the rate of change of current ($\frac{di}{dt}$) through the package interconnect. These two trends are being driven by the increase in IC technology; however, the inductance (L) present in the interconnect is driven by package technology, which is not improving at a rate which would keep $\frac{di}{dt}$ unchanged [11].

When a voltage is induced across the power supply interconnect, the voltage on the chip will be different from the voltage on the system PCB. When delivering a power supply voltage (V_{DD}) to a CMOS device, there will be a voltage drop between the system PCB and the integrated circuit. This phenomenon is called *supply bounce*. Similarly, when the ground return current for a CMOS device traverses an inductive interconnect, a voltage will be induced across the inductive interconnect (causing the voltage at the ground pin of the device (V_{SS}) to be different from the system PCB). This phenomenon is called *ground bounce*. Supply and directly effect the performance of the digital device by increasing its gate delay and causing inadvertent switching [59]. Figure 8.1 shows an ideal CMOS inverter. In this circuit, the on-chip power supplies V_{DD} and V_{SS} are assumed to be the same as on the system PCB. Figure 8.2 shows a CMOS inverter but models the inductance in the supply and ground path. In this case, the voltage from Eq. 8.1 (induced across the inductive interconnect) results in a voltage difference between the supply nodes of the device ($V_{DD\text{–}IC}$, $V_{SS\text{–}IC}$) and the supply voltages on the system PCB ($V_{DD\text{–}System}$, $V_{SS\text{–}System}$).

In modern VLSI designs, between 25 to 50% of the total ($\frac{di}{dt}$) can be consumed in the off-chip driver circuitry and constitutes the largest single source of current consumption [11, 40]; however, V_{DD} and V_{SS} pins are often shared across multiple signal pins that are drawing and returning current for their off-chip driver circuitry. Supply pins are shared to reduce the cost of the package since cost is proportional to the size of the package. Therefore, in order to accurately compute the magnitude of the supply and ground bounce we must modify Eq. 8.1 to consider all of the signal

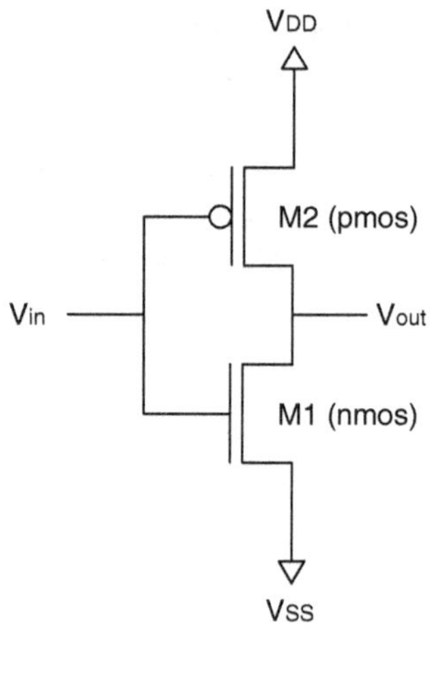

Fig. 8.1 Ideal CMOS inverter circuit

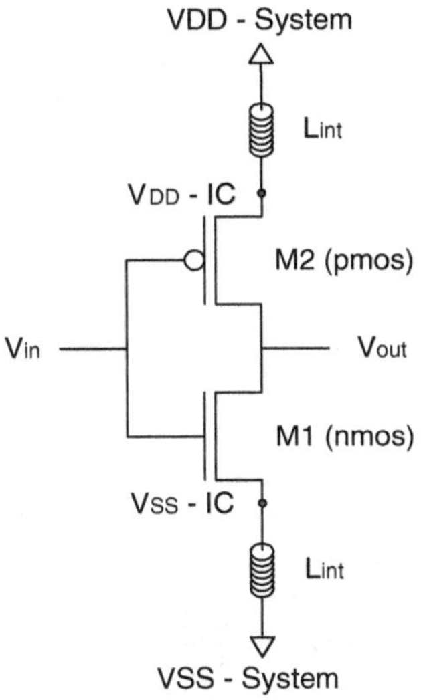

Fig. 8.2 CMOS inverter circuit with supply inductance

pins that are sharing a supply pin to source or return current. For supply bounce the total number of signals (which share a given V_{DD} pin) that are switching from a 0 to a 1 must be considered. Equation 8.2 expresses the total amount of power supply bounce considering all signals that are drawing current through the supply pin inductance. For, the total number of signals that are switching from a 1 to a 0 and are sharing a particular V_{SS} interconnect to return current must be considered. Equation 8.3 expresses the total amount of ground bounce considering all signals that are returning current through the ground pin inductance.

$$V_{Supply\text{–}Bounce} = L_{(V_{DD})} \cdot \sum_{k=0}^{n_{(0\rightarrow 1)}} \left(\frac{di_k}{dt}\right) \tag{8.2}$$

$$V_{Ground\text{–}Bounce} = L_{(V_{SS})} \cdot \sum_{k=0}^{n_{(1\rightarrow 0)}} \left(\frac{di_k}{dt}\right) \tag{8.3}$$

8.2.2 *Inductive Signal Coupling*

When a signal traverses an inductive interconnect in a package it inductively couples to neighboring signals. This is due to the mutual of the magnetic fields for any two current carrying conductors. The magnitude of the mutual inductive coupling voltage between inductive interconnects is governed by the expression:

$$V_M^j = M_{1k} \cdot \left(\frac{di_k}{dt}\right) \tag{8.4}$$

M_{1k} is the mutual inductance between conductor j and conductor k. This relationship can also be expressed in terms of the mutual inductive coupling coefficient K_{ik}. K_{1k} is a dimensionless quantity. Equation 8.5 describes the relationship between M_{1k} and K_{1k}. In this expression, L_j and L_k are the inductances of the two inductively coupled conductors.

$$K_{1k} = \frac{M_{1k}}{\left[\sqrt{L_j \cdot L_k}\,\right]} \tag{8.5}$$

For this work we use the term *victim* for the signal of interest (on which the mutual inductive voltage is induced). We use the term *aggressor* for any neighboring signal that switches and causes a voltage to be induced on the *victim*. Mutual inductive coupling has a cumulative effect in IC packaging. The mutually induced voltage is also dependent on the polarity of the ($\frac{di}{dt}$) in the aggressor's signal path. This means that the total mutual inductive voltage on the victim is the sum of all aggressors' mutually coupled voltage. This unique circumstance creates a situation in which the switching patterns of neighboring signals heavily influence the behavior of the coupled noise on the victim signal. In an extreme case, the aggressors' mutually coupled voltage (of equal and opposite magnitudes) can cancel out and have no effect on the victim. Equation 8.6 describes the total contribution of mutually coupled voltages on a victim signal.

$$V_M^j = \sum_{k=1}^{n} M_{1k} \cdot \left(\frac{di_k}{dt}\right) \tag{8.6}$$

We define the term *Glitch* to describe the situation in which the victim signal is not transitioning and neighboring signal pins are coupling voltage onto it. The *Glitch* voltage caused by switching aggressor signals creates an unwatched voltage on the victim signal. This unwanted voltage can lead to inadvertent switching of digital circuitry that use the victim line as an input.

When the victim signal is transitioning, we use the term *Edge Degradation* to describe the effect of voltage that is coupled from neighboring aggressor signals. Since mutual inductive coupling is cumulative, coupled voltage from neighboring signals can have the effect of aiding, hindering, or not affecting the transition on the victim signal. Figure 8.3 shows the circuit for inductively coupled signal lines in a VLSI package.

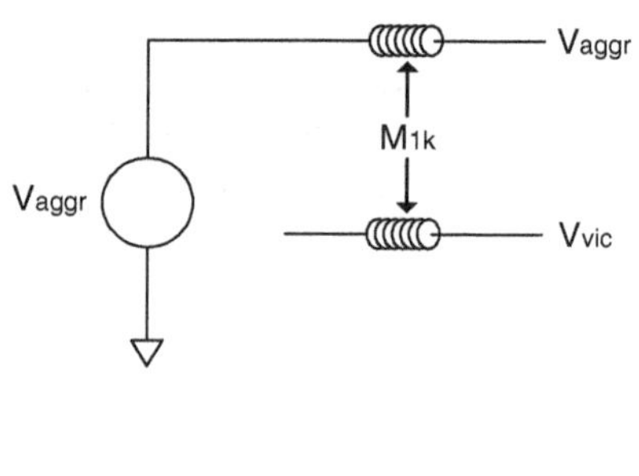

Fig. 8.3 Circuit description of inductive signal coupling

8.2.3 Capacitive Bandwidth Limiting

When capacitance is present along the path that carries signals to and from the devices on the IC, the frequency at which the signal can switch will be reduced. This capacitance will form a low pass RC filter that will prevent frequency components that are above the filters roll-off frequency from passing through the interconnect. This filtering effect will limit the risetime that the package will allow to pass and in turn reduce the overall datarate and performance of the system. The filter that the package creates in the signal path has a standard RC low pass response. The capacitance in the filter comes from the parasitic capacitance present in the package interconnect. The impedance in the filter comes from the characteristic impedance of the system in which the package is placed. Equation 8.7 gives the 10–90% risetime of a single-pole RC filter.

$$t_{rise} = 2.2 \cdot RC = 2.2 \cdot Z_0 \cdot C_{int} \tag{8.7}$$

In this expression Z_0 is the of the system in which the package is placed. C_{int} represents the total capacitance that the package interconnect presents. C_{int} will depend on the parasitic capacitance of the interconnect to ground (C_0) in addition to the parasitic capacitance to neighboring signals (C_{1j}). The total number of neighboring signals that contribute capacitance to C_{int} must be accounted for. Equation 8.8 gives the total amount of capacitance that an individual package interconnect possesses.

$$C^j_{total} = C_0 + \sum_{k=1}^{n} C_{1k} \tag{8.8}$$

The logic level of neighboring signals will affect the amount of capacitance that a victim signal will have to charge or discharge (C_{1k}). C_{1k} is defined as the capacitance between two conductors when one conductor is held at zero potential. Capacitance is defined as $C = Q/V$. This means that the variable voltage that is present on neighboring signals will change the effective capacitance of the victim signal. When a neighboring signal has the same voltage change as the victim signal (i.e., they are transitioning), then the total coupling capacitance between signals will be zero because there is no net charge difference between the two conductors. When a neighboring signal has a static voltage value while transitioning, then the total coupling capacitance between the signals will be C_{1k}. When a neighboring signal transitions in the opposite direction as the victim signal, then the total effective coupling capacitance will be $2 \cdot C_{1k}$ because the voltage excursion that the capacitor undergoes is $2 \cdot V_{DD}$.[1] This situation means that the risetime of a signal will change depending on the logic levels and transitions present on neighboring signals. Figure 8.4 shows the circuit diagram for a signal with capacitive bandwidth limitation.

[1] The voltage across the capacitor is $-V_{DD}$ before the transition and $+V_{DD}$ after the transition (or vice versa).

Fig. 8.4 Circuit description of capacitive bandwidth limiting

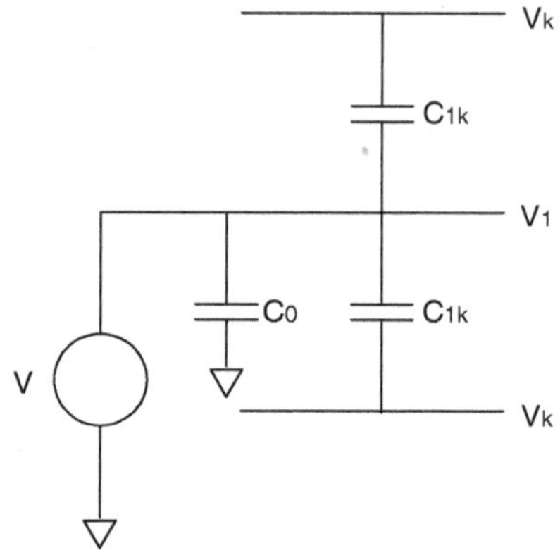

8.2.4 Capacitive Signal Coupling

When a signal traverses a capacitive interconnect in a package, it is electrically modified due to coupling from neighboring signals. This is due to the mutual of the electric fields for any two charge bearing surfaces. The magnitude of the mutual capacitive coupling voltage between capacitive interconnects will depend on the coupling capacitance between the conductors (C_{1k}) in addition to the self-capacitance (C_0) of the victim line. The sum of the coupling capacitance, along with the self-capacitance of the victim line will form a capacitive voltage divider between the victim and aggressor lines. Equation 8.9 describes the voltage magnitude that will be coupled onto a victim line (V_{vic}) from a voltage change on an aggressor line (V_{aggr}).

$$V_{vic} = \left(\frac{C_{1k}}{C_{1k} + C_0}\right) \cdot \left(\Delta V_{aggr}^{k}\right) \tag{8.9}$$

Capacitive coupling can cause unwanted switching of digital circuitry, which limits system performance. As in the case of mutual inductive coupling, the capacitive coupling has a cumulative effect. Glitches on a victim line that occur due to the transitions on neighboring signals can cause unwanted switching of digital circuitry. Also, the cumulative nature of the coupling will lead to an effect that can either aid, hinder, or not effect the victim signal's risetime. Equation 8.10 models the cumulative nature of the capacitive coupling betweens signals. Figure 8.5 shows the circuit for capacitively coupled signal lines in a VLSI package.

$$V_{vic} = \sum_{k=1}^{n} \left(\frac{C_{1k}}{C_{1k} + C_0}\right) \cdot \left(\Delta V_{aggr}^{k}\right) \tag{8.10}$$

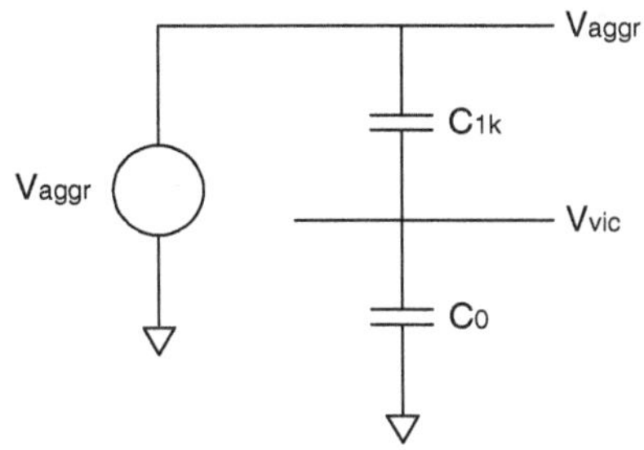

Fig. 8.5 Circuit description of capacitive signal coupling

8.2.5 Impedance Discontinuities

Another source of noise in the package interconnect is reflected energy due to impedance discontinuities. As the risetimes of off-chip driver circuits increase, conductors must be considered as distributed elements. Typically when the interconnect being driven has an electrical length[2] that is longer than 20% of the risetime of the source, then the conductor must be treated as a distributed element and modeled as a transmission line [57]. Current off-chip risetimes are entering into the sub 100ps era. For standard packaging technology this means that a structure longer than 1 to 2 cm must be treated as a distributed element. For these electrical lengths almost all modern packages require the use of transmission line theory for accurate modeling of the package interconnect.

When IC packaging interconnect is treated as a distributed element, becomes critical. When impedance mismatches are present in the package, reflections and risetime degradation will occur. The reflections can lead to unwanted switching of digital circuitry and intersymbol interference (ISI). In addition, risetime degradation will significantly limit system performance. The characteristic impedance of a lossless structure is given by:

$$Z_0 = \sqrt{\frac{L}{C}} \tag{8.11}$$

In this expression the L and C are the inductance and capacitance of a given structure. When considering the impedance of the package, the switching patterns of neighboring signals will alter the inductance and capacitance in Eq. 8.11. Switching signal pins will add or subtract additional amounts of mutual inductance and capacitance to the victim signal. These additional parasitics must be considered in the evaluation of characteristic impedance. This changes the impedance within the package to:

$$Z_0' = \sqrt{\frac{L \pm \sum_1^k M_{1k}}{C \pm \sum_1^k C_{1k}}} \tag{8.12}$$

[2] Propagation delay of the structure.

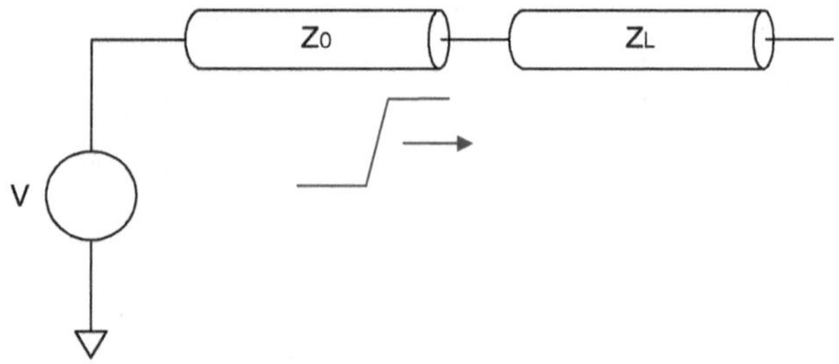

Fig. 8.6 Circuit description of a distributed transmission line

When a wave front is traveling on a conductor with a particular characteristic impedance (Z_0) and encounters a region with a different characteristic impedance (Z_L), reflections will occur. Figure 8.6 shows the circuit diagram for a distributed transmission line.

Γ is the reflection coefficient which represents the magnitude of the reflected voltage (as a fraction of the incident voltage) and is defined as:

$$\Gamma = \frac{Z_L - Z_0}{Z_L + Z_0} \tag{8.13}$$

Γ can be used to find the reflected and transmitted voltage of an incident wave front. Equation 8.14 defines the amount of reflected voltage due to an impedance mismatch. This Equation illustrates the relationship between an impedance mismatch and the magnitude of the reflected voltage. Note that the reflected voltage can have either a positive or negative amplitude depending on whether Z_L is greater than or less than Z_0.

$$V_{reflected} = V_{incident} \cdot \Gamma \tag{8.14}$$

Equation 8.15 defines the amount of transmitted voltage when encountering an impedance mismatch. In this case the magnitude of the reflected voltage will be subtracted from the forward going wave front. This translates into a slower risetime that is propagated through the impedance mismatch. Since risetimes are directly proportional to the datarate that a package can achieve, any reduction in the risetime due to reflected energy will directly effect system performance.

$$V_{transmitted} = V_{incident} \cdot (1 - \Gamma) \tag{8.15}$$

In digital systems typically the system PCB uses a close to 50 Ω. The interconnect structures within the IC package typically do not match to the system PCB impedance and will lead to reflections. When the impedance of a package structure is higher than the characteristic impedance of the system (50 Ω), it is considered an inductive interconnect. It is considered to be inductive because in Eq. 8.12 a higher than optimal inductance will lead to a higher impedance. In the same manner, when the impedance of a package structure is lower than the characteristic impedance of the system, it is considered a capacitive interconnect. Since current packages were originally designed for mechanical considerations, getting a desired impedance is difficult without using expensive and non-standard processes.

As in the case with capacitive bandwidth limitation, switching neighboring pins will alter the net coupling inductance and capacitance that the victim pin experiences. The polarity and quantity of switching signals will alter the effective L and C quantities in Eq. 8.12 and in turn the magnitude of Γ.

8.3 Performance Modeling and Proposed Techniques

All of the noise sources described in Sect. 8.2 can be reduced through aggressive package design. We have seen success in Ball Grid Array (BGA) [51] and Flip-chip packaging [53] technologies that have been able to reduce the electrical parasitics of the interconnect structures within the IC package; however, aggressive package design is often too expensive and unpractical for the majority of VLSI designs. In addition, package design is a slow, evolutionary process that is not predicted to keep pace with the performance increases on the IC. As such, any techniques that can assist in improving performance without moving toward advanced packaging are of considerable value to the VLSI community. In addition, such techniques that model the general RLC parasitics of the package can be used to increase the performance of the package, and extend the useful lifetime of the package.

8.3.1 *Performance Modeling*

In all the noise sources in Sect. 8.2, the amount of noise is ultimately proportional to the rate of change of the current or voltage ($\frac{di}{dt}$ or $\frac{dv}{dt}$). The rate of change of current or voltage can be translated to system performance using standard equations for robust high-speed digital design [57]. Since all of the noise sources are expressed in terms that relate the amount of noise to the rate of change in current or voltage, the acceptable rate of change of voltage or current can be calculated. Using this approach the maximum system performance can be predicted using the electrical package parameters and user-defined noise limits as the two inputs into the model. In this work an analytical model is proposed for system performance which considers each noise source. From this model the maximum rate of change of voltage or current can be found for any given package and set of noise limits. This directly predicts the per-pin datarate and ultimate bus throughput for a given package. This analytical model is also used to predict the performance increase for the various solutions that are presented. The performance model presented in this work is shown to be accurate to within 10% of analog simulator results which are much more computationally expensive.

8.3.2 *Optimal Bus Sizing*

One of the biggest challenges for VLSI designers is selecting the size and configuration for their off-chip busses [40]. The main problem is that simply adding pins to an off-chip bus does not necessarily increase the throughput in a linear manner [49].

As pins are added to a package to increase the bus size and throughput, the noise induced by the signals that are simultaneously switching leads to a significant increase in noise. This means that as signals are added, the per-pin datarate of each pin needs to be decreased to meet the system noise limits. As a result, the overall throughput of an off-chip bus approaches an asymptotic limit as signals are added. As a consequence, multiple bus configurations can meet the same throughput objective. In order to aid in the selection of the optimal bus configuration, the cost of the bus is considered in the model. By including the cost of the signal, power, and ground pins that are needed to expand the size of a bus, the most cost-efficient bus configuration can be found. This allows designers to quickly compare two bus configurations of equal throughput and choose the least expensive solution. The model considers the electrical parameters of the package in addition to the per-pin cost, which allows the model to compare configurations in different packages. The metric of *Bandwidth per Cost* is defined as a means to find the most cost-effective bus configuration. The results presented in this work are supported by the industry trend of moving toward faster narrower busses over slower wider busses to achieve the same throughput at a lower cost.

8.3.3 Bus Encoding

When determining the maximum performance of a bus the worst-case noise limits must be considered. In all of the noise sources listed in Sect. 8.2, a worst-case bus pattern is assumed to be present on the bus. In the case of supply bounce all of the signals in the bus are assumed to be switching in the same direction. In the case of bandwidth limitation and signal coupling, the middle signal in the bus is either switching in one direction while all of the other signals on the bus are switching in the opposite direction, (Edge Degradation) or the middle signal is static while all other signals are switching in the same direction (Glitch). This work presents two bus encoding techniques that encode the off-chip data prior to leaving the IC. The encoding ensures that all patterns that result in noise of a magnitude greater than a user-specified limit are eliminated. By doing this a lower level of noise in the system is achieved which translates into increased performance. The techniques presented in this work show that the performance of the bus is increased dramatically *even after* considering the overhead of the encoder algorithm.

The first encoding technique is called *Bus Expansion* which maps the original set of logic vectors on the IC into an expanded set of vectors prior to leaving the IC. This expanded set of vectors ensures that noise patterns of a magnitude greater than the user-specified limit are never present on the data that is traversing the package.

The second encoding technique is called *Bus Stuttering*, which uses the original number of package pins but avoids the worst-case noise patterns by inserting intermediate logic vectors between outgoing pairs of vectors that induce noise with a magnitude greater than the user-specified limit. This approach allows the algorithm to be tailored for packages that are dominated by interconnect that is inductive, capacitive, or both. Experimental results show that the encoder circuits can be integrated into a modern CMOS process and achieve a minimal delay and area impact on the design. The encoders are shown to improve performance of off-chip busses up to 225% by avoiding data patterns which result in noise violations.

8.3.4 Impedance Compensation

Impedance discontinuities are caused by excess inductance or capacitance that is present in the package interconnect. The term *excess* refers to any inductance or capacitance that causes the interconnect impedance to be different from the system impedance. If the interconnect has a higher impedance than the system ($Z_L > Z_0$), then the interconnect is said to have excess inductance. If the interconnect has a lower impedance than the system ($Z_L < Z_0$), then the interconnect is said to have excess capacitance. When this occurs, the impedance discontinuities cause reflections which limit the performance of the system. To address this problem a compensation technique is presented that adds capacitance or inductance near the interconnect to match its impedance to the system impedance. If the spatial location of the compensation inductance/capacitance is near the interconnect, then the interconnect and compensation will be seen as a lumped element and will be governed by Eq. 8.12. By using this compensation technique, a forward traveling wave front will see the compensated package interconnect as a matched impedance and reflections will be avoided.

Two compensation techniques are presented in this work. The first technique is a *static compensation* approach in which the pre-defined compensation elements are placed on the package and IC substrate. This technique surrounds the interconnect with the appropriate inductance or capacitance to achieve the impedance match.

The second technique is a *dynamic compensation* approach in which programmable compensation elements are placed on-chip. By placing the compensation on-chip, the interconnect impedance can be adjusted after the IC is packaged. This has the advantage of being able to accommodate any design or manufacturing tolerances that may vary the interconnect's original impedance. Both techniques are shown to be able to compensate for all reasonable ranges of package interconnect impedance and significantly reduce reflections. The compensation techniques are shown to reduce reflected noise as much as 400% for broadband frequencies up to 3 GHz.

Figure 8.7 show a summary of the improvement in bus performance that can be achieved using the techniques proposed in this work. The original performance is based on a bus size of 4 bits with one V_{DD} and one V_{SS} pin where supply bounce is the failure mechanism. The performance is improved by encoding the data to avoid data patterns which result in noise limit violations and in turn limit the maximum throughput that can be achieved. The performance improvement considers the overhead of the encoder. For all package technologies studied in this work, the supply bounce due to the inductive nature of the interconnect is the limiting factor. Moving toward advanced packaging reduces the parasitic inductance present in the package which results in faster performance. In addition, when the parasitics of the package are reduced, the improvement techniques presented in this work will have more of a positive impact.

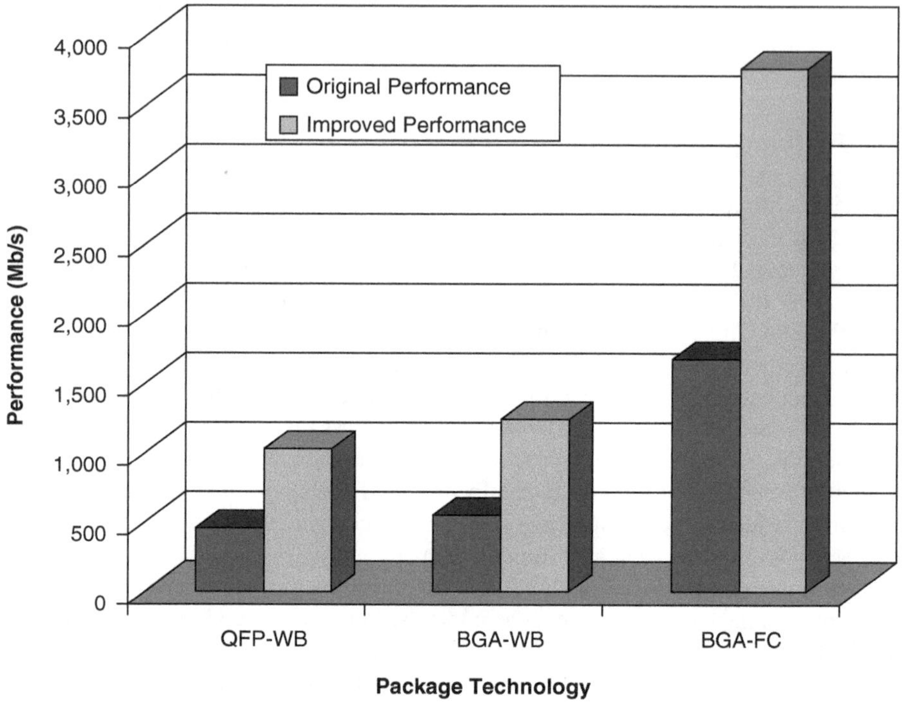

Fig. 8.7 Performance improvement using proposed techniques

8.4 Advantages Over Prior Techniques

This section presents previous research and current techniques on the topic of package performance modeling and noise reduction. In each area the contribution of this work is compared and contrasted.

8.4.1 Performance Modeling

The most commonly used technique to predict the performance of an IC package is through analog simulation tools such as SPICE [2, 76]. SPICE (**S**imulation **P**rogram for the **I**ntegration **C**ircuit **E**nvironment) is a general purpose circuit simulator. SPICE simulates circuits by simultaneously solving the governing electrical equations for each element in the circuit. By doing this the detailed analog behavior of each node within the circuit can be monitored in either the time or frequency domain. While SPICE is a powerful tool for relatively small circuits, it is typically too computationally expensive to apply to entire VLSI designs [47].

The analytical model presented in this work forms a simple expression that relates a given noise source magnitude to a rate of change in current or voltage ($\frac{di}{dt}$ or $\frac{dv}{dt}$). The noise source magnitude is proportional to the electrical parasitics of the IC package in addition to the rate of change in current or voltage. The rate of change of current or voltage is related to overall system performance by applying a series of design rules for a robust digital system. Using this framework the acceptable noise limits for any noise source can be converted into predicted system performance metric for a given package. This technique is linear resulting in a dramatically lower computation time compared with SPICE with high accuracy. This enables the evaluation of more bus configurations, packages, and noise limits which enhances the decision-making ability of the VLSI designer. In addition, integration of this approach into digital Computer Aided Design (CAD) tools becomes more practical.

8.4.2 Optimal Bus Sizing

Traditionally, inter-chip communication is performed using wide parallel busses. The standard approach to achieving the desired system bandwidth is to increase the number of pins on the package until the desired throughput is attained. There are three main problems with this approach:

- *Cost of packaging.* Package cost scales faster than linearly with the number of I/O pins that are needed, and accounts for a large contribution to the overall chip price [40].
- *Performance.* Wide parallel busses experience a host of signal integrity issues associated with simultaneous switching of digital signals [23, 47, 66] as outlined in Sect. 8.2. The noise induced in the package is proportional to the number of off-chip signals that are switching simultaneously. One solution to this problem is to increase the number of power and ground pins in the bus configuration. This reduces the inductance in the power supply current path in addition to reducing the magnitude of the coupled noise between signals; however this increases the cost of the package because the number of I/O pins increases. Another solution to the package parasitic problem is to move toward advanced packaging technologies such as flip-chip bumping to reduce the parasitic inductance and capacitance in the package [53]. While this solution does reduce the noise in the package, advanced package technologies dramatically increases the price of the IC.
- The increases in package bandwidth do not scale at the same rate as on-chip core frequencies [47]. The traditional approach of widening parallel busses to match the inner core's datarate is impractical not only from a cost viewpoint, but also because the signal integrity problems mentioned above limit how wide busses can be. The paradox of the wide parallel bus is—intuitively, adding I/O should produce a linear increase in system throughput but in reality suffers an asymptotic limit due to additional noise sources that arise as more signals switch. Parallel busses have to be ran at lower speeds as their width increases, which inherently limits their throughput.

Recently we have seen the emergence of narrower busses that run at higher per-pin datarates [23, 47]. These new busses include Rapid I/O [82], PCI Express [77], and

Hyper Transport [30]. All of these busses take advantage of the fact that the same bandwidth can be achieved by using less signals that run faster than a traditional wider parallel bus that uses more signals at a reduced per-pin datarate. By using narrower busses, the desired throughput of the bus is achieved at a lower cost on account of using fewer package pins. Regardless of whether the inter-chip communication uses a slower, parallel bus or a faster, narrow bus, the objective is the same: the inter-chip bus must deliver the highest throughput in the most cost-effective manner. This is a challenging problem due to the faster than linear increase in the cost of adding I/O pins that must be balanced with the fact that there exists an asymptotic limit to how much bandwidth can be attained by widening the bus.

Much work has also been done to increase the throughput of the IC package by moving toward advanced packaging. Research has been performed on both the level 1 (IC to Package Substrate) [53, 70] and level 2 (Package to System PCB) [51, 52, 72] connections with the ultimate goal of reducing the parasitic inductance and capacitance of the interconnect. While these approaches increases performance, they are typically too expensive for the majority of VLSI designs.

Recently there has also been considerable research in the design and characterization of low signal count busses [24, 31]. This approach targets busses that use low channel counts, which run at faster datarates than traditional IC package pins have historically been able to achieve. The main goal of this thread of research has been to increase the performance of an individual off-chip channel.

The bus sizing technique presented in this work aims at finding the optimal configuration of signal, power, and ground pins that yields the most system bandwidth at the least amount of cost. The metric of *Bandwidth per Cost* is introduced which gives VLSI designers a quick method to compare different bus configurations for cost-effectiveness.

8.4.3 Bus Encoding

There has been work done in the area of reducing package noise by altering the signals prior to leaving the IC. Pipeline Damping was presented in [79], with the aim of reducing the total $\frac{di}{dt}$ by implementing a multi-valued output circuit. This work reduced the average $\frac{di}{dt}$ in the output stage by limiting the output swing of the driver. While this technique reported significant reduction in average off-chip noise, the peak noise limit of the output stage was not reduced which resulted in the same off-chip datarate.

Other techniques aimed specifically at the design of the output stage to limit the net ground bounce in the package [73]. These techniques were successful in reducing ground bounce; however, other noise sources were not addressed.

On-chip bus coding techniques have been applied to avoid capacitive cross-talk in long busses [26, 35, 36, 100]. These approaches encodes the data prior to driving it on the bus and remove the worst-case capacitive cross-talk patterns. These techniques have been shown to increase the performance of on-chip busses by reducing the interconnect delay. The work in this monograph uses a similar approach but addresses the multiple noise sources present in the inductive off-chip interconnect.

Statistical encoding techniques have been used in audio and video applications to aid in relieving network congestion [48, 92]. These techniques have been successfully applied to MPEG and DVD protocols [56, 83]. The main approach of these encoding algorithms has been to eliminate similar or redundant data packets that occur sequentially on the network. This allows a comparable quality of audio/video (AV) without needing to send each and every data packet associated with the data compression algorithm.

The contribution of the encoding techniques presented in this work differ from previous techniques in that they specifically address the electrical noise sources in the physical interconnect. This makes them ideal for reducing noise within IC packaging. It also makes them suitable for any application in which the switching patterns directly effect performance such as on-chip capacitive busses, power minimization, or in statistical A/V protocols.

8.4.4 *Impedance Compensation*

Much work has been done in the characterization of the electrical parameters of package interconnect [69, 74, 102]. Previous work has demonstrated the severe impedance mismatch that occurs due to the interconnect structures. Altering the impedance of an electrical structure in the package has typically been difficult since adding additional capacitance or inductance near the structure has been impractical due to the relatively large size of the component. The standard approach has been to tolerate the resultant reflections from the package. The magnitude of the reflections was one of the limiting factors to a package's performance. This approach is no longer practical as the risetimes of digital signals continue to increase. Many improvements have been made in integrating on-chip and on-package capacitors and inductors [46, 63, 64]. New materials have allowed placing impedance altering components near the package interconnect, which has enabled the possibility of matching the impedance of the interconnect to the system impedance [13, 16, 80]. This work uses the advances in on-chip and on-package componentry to construct a technique that alters the impedance of the package interconnect [27]. This technique allows a better impedance match through the package, which results in reduced reflections and higher system throughput.

8.5 Broader Impact of This Monograph

This monograph presents techniques that specifically model and improve performance in IC packages. Since all of the modeling and performance techniques are described using a common mathematical framework, the work in this monograph can be easily applied to a wide variety of electronic applications. The encoding techniques are formulated to eliminate data sequences which result in unwanted noise, which makes them suited for application in power reduction, audio/video compression, and internet fabric congestion. In addition, since all the techniques presented in this work are constructed to alleviate the detrimental effects of the inductance and capacitance of the interconnect, this naturally makes them applicable to any noise-prone electrical interconnect in a digital system. These applications include connectors, backplanes, and cables. Since all of the techniques are formulated independent of technology or process, they can be easily implemented as *intellectual property* cores for use in a wide variety of VLSI designs and implementation in Field Programmable Gate Arrays.

8.6 Organization of Part II of this Monograph

Part II of this monograph is organized as follows: Chapter 9 describes the package technology that is studied in this work. The construction of three industry standard packages are presented. The three packages are the *Quad Flat Pack* (QFP) with *Wire Bonding* (WB); the *Ball Grid Array* (BGA) with *Wire Bonding*; and the BGA with *Flip-Chip Bumping* (FC). In the past decade the QFP-WB package has been the most popular style of packaging. The BGA-WB package is presently the most popular package while the BGA-FC package is predicted to be the most common package in the next decade. Packages construction as well as electrical parameters are presented for all three packages. Chapter 10 presents the terminology and background information that will be used throughout the rest of the monograph. Chapter 11 describes the analytical model for performance and noise level prediction. Chapter 12 presents the optimal bus sizing technique to determine the most cost-effective bus configuration. Chapters 13 and 14 present the two bus encoding techniques that reduce package noise by avoiding switching patterns that result in noise that is greater than a user-specified limit. Chapter 15 presents the impedance compensator methodology. Chapter 16 discusses future trends in VLSI and how this research can be applied to increase digital system performance. Finally, conclusions are drawn in Chap. 17.

Chapter 9
Package Construction and Electrical Modeling

The construction of an IC must accomplish many tasks. The first is to electrically connect the signals on the IC to the signal paths on the system PCB. The second is to electrically conduct power from the system PCB to the devices on the IC. In addition, the package must provide mechanical protection for the IC as well as thermal dissipation. The typically consists of two levels of interconnect. The first is the connection from the IC to the package (level 1) and the second is from the package to the system PCB (level 2).

9.1 Level 1 Interconnect

The level 1 interconnect electrically connects the signals and power on the IC to the package. On the IC the transistors reside on the lowermost layers of the die, above the substrate. The various metal layers contain the electrical interconnect which allows the construction of complex digital circuitry and power distribution grids which supply current to the transistors [62]. The highest metal layer contains the pads to which the level 1 interconnect will contact. In a typical IC, the metal layers utilize Aluminum or Copper [96]. For a modern VLSI IC, the typical sizes of the pads can range from $35 \times 35\,\mu m$ to $100 \times 100\,\mu m$.

9.1.1 Wire Bonding

One of the most common and proven level 1 interconnect structures is the *Wire Bond*. A can be constructed with Gold, Aluminum, or Copper and is a thin round conducting wire that bonds to the pads on the IC and on the package. The wire bond can range in diameter from $7\,\mu m$ to 1 mm depending on the technology of the automated bonding machine. The connection between the wire and the pad is accomplished using thermosonic energy [96] which anneals the two metals into a robust mechanical joint. The contact to the IC is accomplished using a *ball bond* connection. A ball bond is formed when the automated bonding machine approaches the pad in a perpendicular manner. This connection requires a square pad on the IC

C. Duan et al., *On and Off-Chip Crosstalk Avoidance in VLSI Design*,
DOI 10.1007/978-1-4419-0947-3_9,

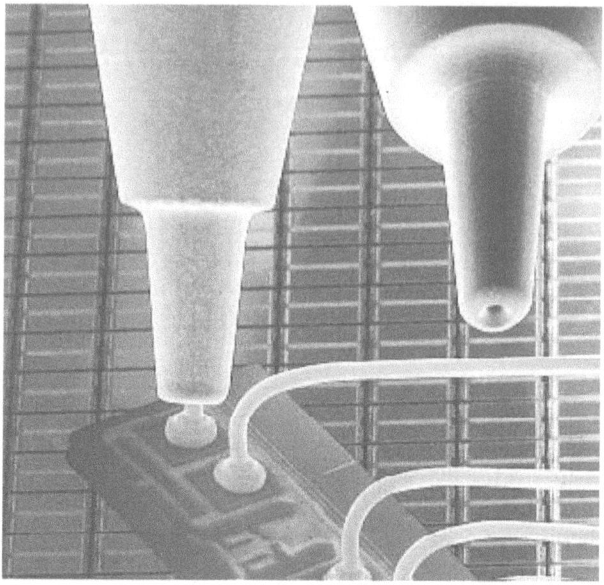

Fig. 9.1 SEM photograph of ball bond connection

which minimizes area on the die. Figure 9.1 shows a Scanning Electron Microscope (SEM) photograph of a ball bond joint [37].

The connection between the wire bond and the package pad can be formed using either a ball bond or a *wedge bond.* A wedge bond is accomplished when the automated bonding machine approaches the pad in a parallel sweeping manner. This type of connection requires a rectangular pad on the package, typically on the order of $100 \times 400\,\mu m$. The wedge bond is used as a way to increase the speed of the automated bonding process; however, when the area on the package is a constraint, a ball bond can also be used which reduces the package pad size by using a square shaped pad. Figure 9.2 shows an SEM photograph of a wedge bond joint [37].

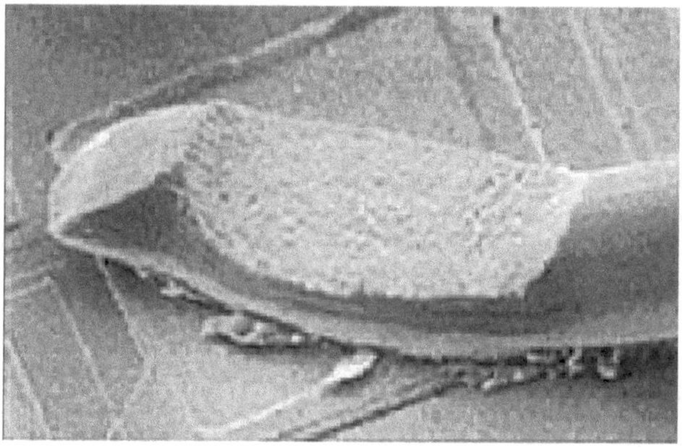

Fig. 9.2 SEM photograph of wedge bond connection

Fig. 9.3 SEM photograph of a wire bonded system

The automated wire bonding process has been refined over the past decades to increase joint strength and reduce process time. Modern wire bonding machines are capable of making thousands of individual connections per minute. The increased speed of the bonding process reduces the overall cost of the package. The low cost and reliability of the wire bond has made it the most popular choice for the level 1 interconnect within an IC package. Figure 9.3 shows an SEM photograph of the entire wire bond structure for an IC substrate with pads placed on the perimeter of the die [96].

While the wire bond interconnect is the most popular level 1 interconnect due to its low cost and mechanical robustness, its electrical parasitics can be considerable, especially when running at today's data rates. The wire bond assembly process produces an interconnect structure that is separated from its return current path by a relatively large distance. This has many electrical disadvantages. The first is that the wire bond contains a significant amount of self-inductance (L_{11}) due to the loop area of the return current path. This leads to power supply bounce (Eqs. 8.2 and 8.3), mutual inductive signal coupling (Eq. 8.6), and inductive impedance discontinuities (Eq. 8.13). The second disadvantage is that since the wire bond is far from its return path, its self capacitance (C_0) is reduced and has a similar value as the magnitude as the mutual capacitance (C_{1k}) to other signal wires. This leads to increased capacitive signal coupling (Eq. 8.9) and capacitive bandwidth limiting (Eq. 8.4). While these electrical factors limit the performance of wire bonded packages, wire bonding is still used in more than 95% of all VLSI design starts due to its reliability and cost-effectiveness [11, 40, 96].

9.1.2 Flip-Chip Bumping

To address the electrical performance limitations of wire bonding, *Flip-Chip bumping* was introduced. In square pads are placed on the topmost metal layer of the IC

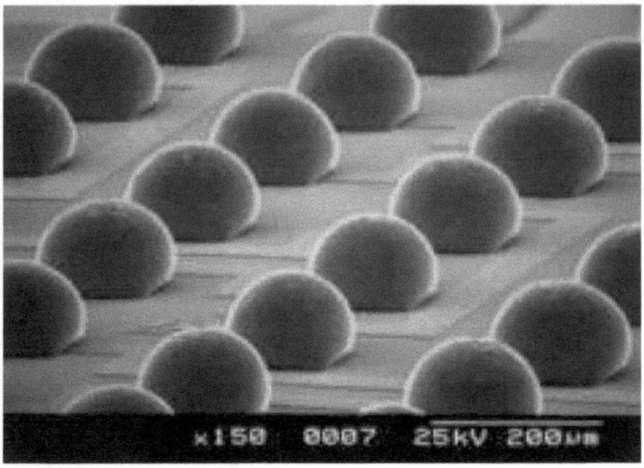

Fig. 9.4 SEM photograph of flip-chip bump array

substrate. Upon these pads, a soldering compound is placed either through chemical vapor deposition (CVD) or plating. The solder material is then heated to a temperature in which it melts or *reflows*. When the solder material becomes molten, the surface tension of the solder attempts to reduce its surface area and forms a spherical object. Since the bumping is accomplished through a chemical process instead of a mechanical process (as in wire bonding), the IC substrate pads can be arranged in an array pattern and can be much smaller. This allows the entire surface of the IC to be used for bumping pads, which dramatically increases the number of level 1 interconnects that are possible for a given die size. Figure 9.4 shows an SEM photo of flip-chip bumps after reflow [37].

The package substrate contains a complimentary array of pads that matches the array pattern on the IC substrate. The IC substrate is then turned face-down and placed on the package substrate. The IC and package substrates are then subjected to a reflow process in which the solder bumps once again become molten and form an electrical and mechanical connection between the IC and the package pads. The surface tension of the solder bumps prevents the weight of the IC substrate from completely collapsing the solder bump spheres. This leaves an air gap between the IC substrate and the package substrate everywhere there is no flip-chip bump. This gap is then filled with a non-conductive adhesive using an *underfill* process. The underfill process creates a solid mechanical structure that absorbs the thermal expansion mismatches between the IC substrate and package substrate. The underfill process also prevents moisture and contaminants from getting into the bump array, which can cause local thermal expansion mismatches. The underfill process is critical due to the extreme temperature range that an IC will cover during normal operation. A large thermal expansion mismatch can lead to stress fractures in the flip-chip bumps which could sever the electrical connection from the IC to the package.

Electrically, flip-chip bumping has many advantages. First, the flip-chip bumps are significantly smaller than bonding wires due to the use of a chemical process instead of a mechanical process to form the connection. Flip-chip pads can be as small as 35 μm in diameter compared to 100 × 100 μm for a wire bond connection. The reduced size of the interconnect leads to reduced parasitic inductance and capacitance associated with the connection. The lower electrical parasitics reduce all of the noise sources described in Sect. 8.2. The second electrical advantage of flip-chip bumping is that the pads can be placed in an array pattern. This allows more pads to be placed for a given substrate size compared to traditional perimeter placement used in wire bonding. The increased pad count leads to higher signal density and the possibility of large amounts of redundant power and ground connections. When redundant power and ground paths are used it reduces the overall inductance in the power distribution path. This results in less supply bounce (Eqs. 8.2 and 8.3).

The main disadvantage of flip-chip bumping is that it is more expensive than wire bonding. The cost is higher due to the manufacturing process that is required for flip-chip bumping. The bumping process must be carried out at the wafer level prior to individual IC separation. This means that ICs that have been identified as failures during electrical wafer test still receive the bumping process. As a result, more time is associated with the process, which results in greater cost. In addition, the bumping equipment must be able to accommodate the entire wafer instead of just the individual die as in wire bonding. As wafer sizes continue to increase, the bumping equipment must be continually upgraded. These manufacturing factors have prevented the wide spread adoption of flip-chip bumping despite its electrical advantages.

9.2 Level 2 Interconnect

The level 2 interconnect electrically connects the signals and power on the package to the system PCB. The system PCB is constructed using a lamination process in which the outermost layers contain surface mount (SMT) pads, to which the package is soldered. Since a lamination process is used, the pads on the PCB are much larger than that used in the level 1 connection. There are two main styles of level 2 interconnect; lead frame and a ball grid array.

9.2.1 Lead Frame

A is a structure that forms the electrical and mechanical connection between the package and the system PCB. The lead frame begins as a single piece of copper-based alloy sheet metal that contains a die mounting paddle and lead fingers. The features of the lead frame are formed by stamping or etching, which is then followed by a plating finish to reduce oxidation. The die mounting paddle is used to support the die during assembly. The lead fingers are used to form the connection to the system PCB and extend outwards from the perimeter of the die paddle. Lead frames only support wire bonding connections from the die. During assembly, the die resides on the die paddle while the wire bonding takes place to the outward extending lead fingers. The die paddle is then removed and the die is covered with a non-conductive encapsulant that protects the IC substrate and helps dissipate heat. The lead fingers

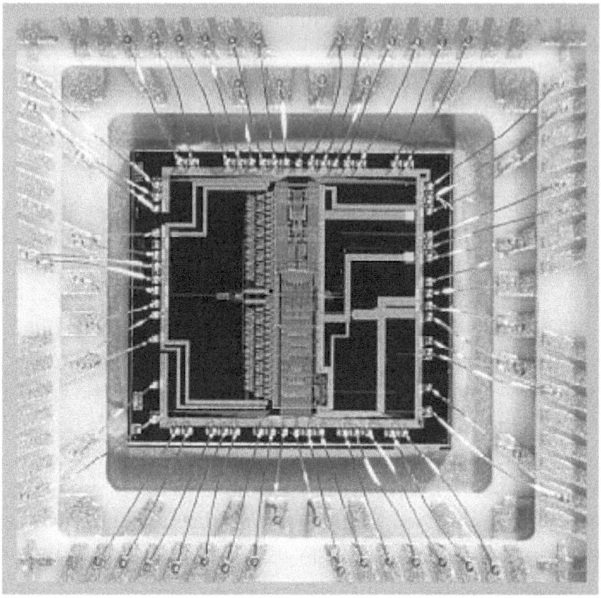

Fig. 9.5 Lead frame connection to IC substrate

extend out of the encapsulant and are then trimmed and formed to their final shape prior to mounting to the system PCB. At this point the leads can be shaped into through-hole or SMT connections. Figure 9.5 shows the wire bonding from the IC substrate to the fingers of a lead frame [37].

The lead frame assembly process has been refined over the past decades to yield an inexpensive solution to IC packaging; however, the lead frame suffers from the same electrical disadvantages as the wire bond in that the interconnect is relatively long which creates a large return current path. The large return current path increases the self-inductance of the lead and the mutual coupling capacitive between leads, both of which increase package noise that limits the lead frame's performance. Another drawback of the lead frame is that it limits signal from the package to a perimeter egress. This limits the overall signal and power density that can be achieved by this style of interconnect.

9.2.2 Array Pattern

To improve upon the lead frame package interconnect, packaging was introduced. BGA technology is very similar to flip-chip bumping except that the solder balls reside between the package and the system PCB. Since the laminated construction of the PCB limits the feature sizes that can be achieved, the density of the array pattern that can be achieved is much less than flip-chip bumping. Current BGA packages range from 0.5 to 1.27 mm contact pitch [96]. Since the level 2 solder balls are much larger than the flip-chip bumps, an underfill encapsulant is not needed since

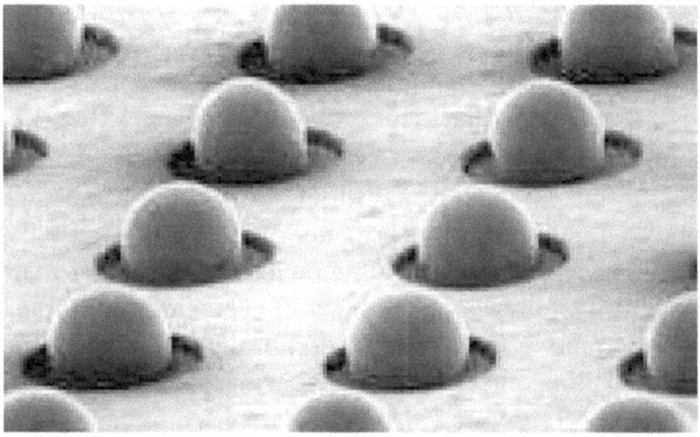

Fig. 9.6 SEM photograph of the bottom of a 1 mm pitch BGA package

Fig. 9.7 Photograph of a 1 mm pitch BGA package

the solder balls can absorb the majority of the thermal expansion mismatch between the package and the system PCB. In addition, the BGA package substrate is typically implemented using a PCB so the thermal expansion difference between the package and the system PCB is minimized. As with the flip-chip bumping interconnect, the BGA interconnect reduces the overall parasitic inductance and capacitance in the interconnect by reducing the electrical length relative to the lead frame approach. Also, the array configuration of the interconnect leads to the easy implementation of redundant power and ground pins which decreases the self-inductance in the power supply path. Figure 9.6 shows an SEM photograph of the underside of a 1 mm pitch BGA package [37].

Figure 9.7 shows a photograph of an entire BGA package highlighting the array configuration of the interconnect [37].

9.3 Modern Packages

This work will focus on three of the most common packages used in modern VLSI designs. These packages represent the past, present, and future technologies of IC packaging and illustrate the electrical, mechanical, and cost trade-offs that have been made by VLSI designers.

9.3.1 *Quad Flat Pack with Wire Bonding*

One of the most common packages over the past decade has been the with Wire Bonding (QFP-WB). The QFP-WB uses wire bonding as its level 1 interconnect with ball bonds on the IC substrate and wedge bonds on the lead frame. The QFP lead frame extends from all four sides of the IC substrate, which allows pads to be placed along the entire perimeter of the IC. The QFP has been extremely successful due to its ability to use a common lead frame across many sizes of IC substrates. The flexibility of the wire bond interconnect allows multiple sizes of dies to be connected to the frame. The standardization of the lead frame size allows for the optimization of the plating, encapsulating, and trimming process which has driven the cost out of this package. Figure 9.8 shows a cross-section and top view of a QFP-WB package.

The QFP-WB package is mounted to the system PCB using a pattern of perimeter placed pads which align to the lead frame contacts of the package. This package is mounted on the surface of the target PCB which allows placement on both the top and bottom of the system PCB. Figure 9.9 shows a cross-section of the entire assembly of a QFP-WB package to a system PCB. Signals and power make contact to the inner layers of the system PCB using *vias*. While the QFP-WB is a cost-effective package, the electrical limitations in both the level 1 and level 2 interconnect has slowed the use of the QFP-WB package.

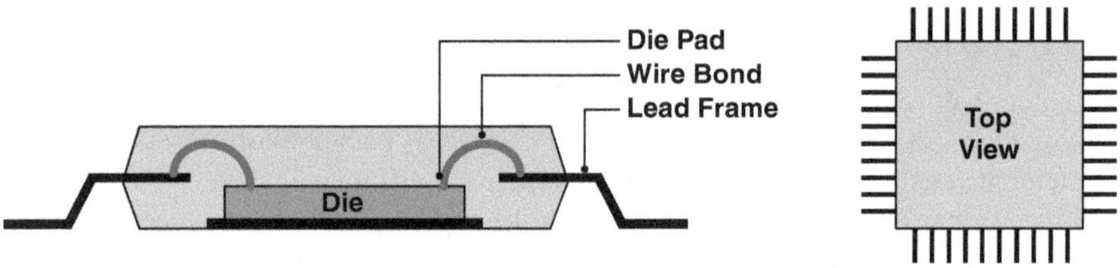

Fig. 9.8 Cross-section of quad flat pack with wire bonding package

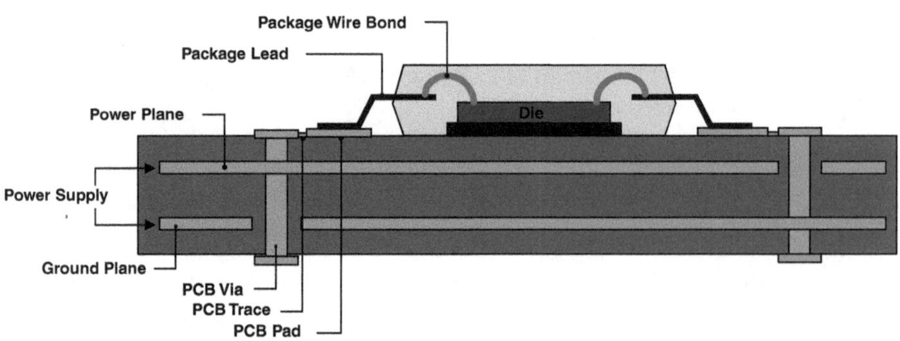

Fig. 9.9 Cross-section of system with QFP wire bond package

9.3.2 *Ball Grid Array with Wire Bonding*

The most popular package in use today is the Ball Grid Array with Wire Bonding (BGA-WB). This type of technology uses a PCB substrate within the package. The level 1 interconnect is implemented with wire bonds that connect the ball bond pads on the IC to the wedge bond pads on the package PCB. The level 2 interconnect is implemented using an array of solder balls on the underside of the package PCB. The construction of the BGA-WB allows either a partial or fully populated array of pads on the bottom side of the package. Figure 9.10 shows the cross-section of a BGA-WB package and the bottom side array of solder balls.

The BGA-WB package addresses the electrical parasitic problem in the level 2 interconnect by moving away from the lead frame approach. This shift in technology has allowed much higher performance in this style of package. In addition, the array configuration of the solder balls allows much higher pin counts to be realized in the same system PCB area. The BGA-WB is mounted to the system PCB using a reflow process, which creates the bond between the package pads and the system PCB pads through the solder ball. Signals and power of this package make contact to the inner layers of the system PCB using vias. Figure 9.11 shows the cross-section of a BGA-WB package that is mounted to a system PCB.

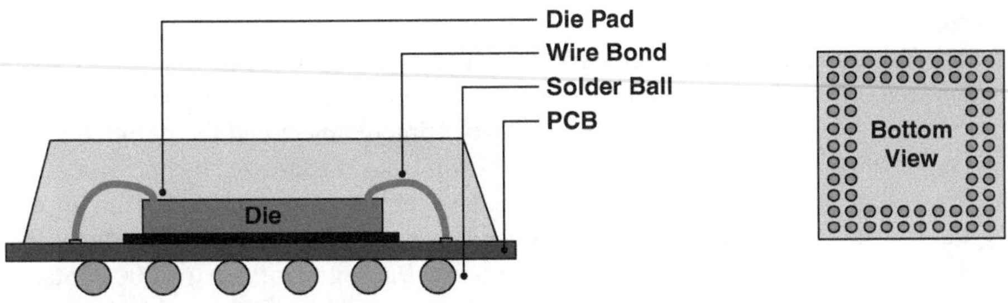

Fig. 9.10 Cross-section of ball grid array with wire bonding package

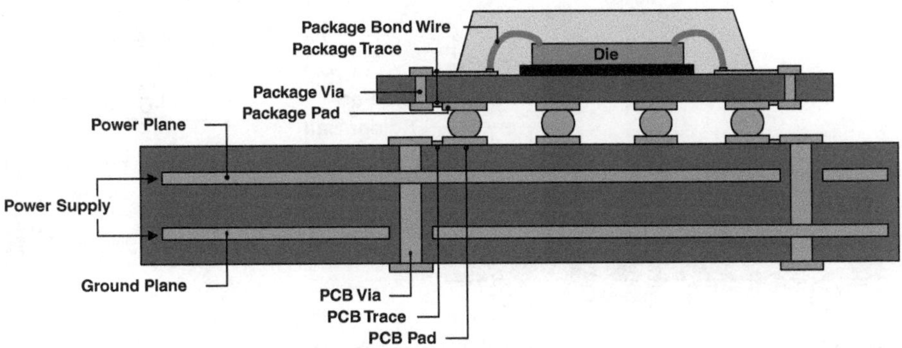

Fig. 9.11 Cross-section of system with BGA wire bond package

The BGA-WB has gained wide adoption due to its combination of the inexpensive level 1 wire bonding interconnect with the advantages of an array style level 2 interconnect. The main drawback of the BGA-WB is that it still suffers the electrical parasitics associated with the level 1 wire bond interconnect. The wire bond interconnect also restricts the pad placement on the IC to a perimeter pattern which inherently limits the number of level 1 connections. This is the main limitation to the number of signals and power contacts that can be brought out to the system PCB. Despite these disadvantages, the BGA wire bond is the most widely used package for current VLSI design starts due to its electrical advantages over the QFP-WB and its cost-effectiveness relative to non-wire bonded packages.

9.3.3 Ball Grid Array with Flip-Chip Bumping

The next step in the evolution of VLSI packaging is the Ball Grid Array with Flip-Chip bumping (BGA-FC). This package addresses the electrical parasitics of the wire bond by implementing flip-chip technology as its level 1 connection. When combined with the electrical improvements gained by using a level 2 BGA interconnect, the BGA-FC promises to meet the needs of VLSI systems into the next decade. Figure 9.12 shows the cross-section of a BGA-FC package and the bottom side array of solder balls.

The BGA-FC package can achieve very high signal counts due to its array style pad pattern. In a BGA-WB package the total number of interconnects is limited by the wire bond noise that exists in the level 1 interconnect and the perimeter pad placement on the die. In the BGA-FC package this issue is addressed by implementing an array style interconnect between the IC and the package substrate. This allows higher signal counts and redundant power and ground connections which reduces the package noise and increases system performance. Packages with up to 2000 contacts have been successfully implemented using ball grid array with flip-chip bumping technology [40]. Figure 9.13 shows the cross-section of a BGA-FC package that is mounted to a system PCB.

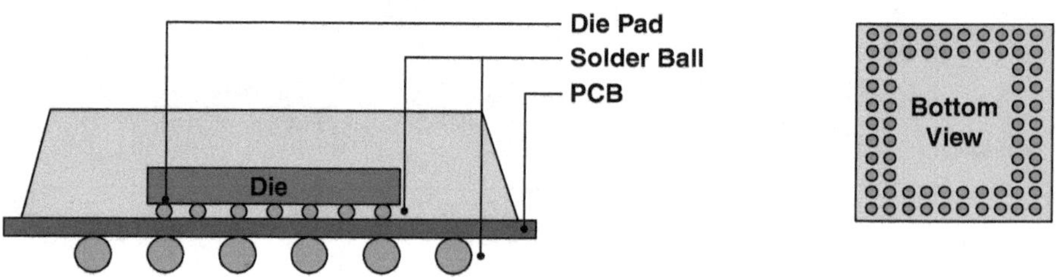

Fig. 9.12 Cross-section of ball grid array with flip-chip package

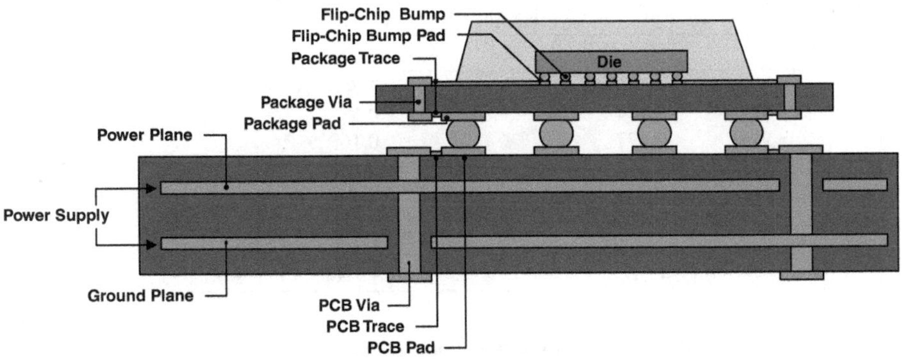

Fig. 9.13 Cross-section of system with BGA flip-chip package

9.4 Electrical Modeling

This section presents the results of the electrical parameter extraction performed to determine the inductance and capacitance magnitudes for the three packages studied in this work. The three dimensional structures within the package were modeled and simulated using electromagnetic field solvers. A combination of *Raphael* from *Avant!* [15] and *Advanced Design System* from *Agilent Technologies, Inc* [50] were used to accomplish the parameter extraction.

9.4.1 Quad Flat Pack with Wire Bonding

The QFP-WB package studied in this work uses a 100-pin lead frame with 25 leads per side on 0.8 mm pitch. The lead frame is composed of a copper based alloy with a tin plated finish. The IC is implemented on a 5 × 5 mm silicon die. The die is connected to the lead frame with 25 μm diameter gold wire bonds. The bonding wires attach to the lead frame using wedge bonds onto 100 × 400 μm pads. The wire connections to the IC are formed using ball bonds onto 100 × 100 μm pads. The wire bonds range is length from 3 to 5 mm.

9.4.2 Ball Grid Array with Wire Bonding

The BGA-WB package studied in this work is a 20 × 20 array of 1 mm pitch, fully populated with solder balls. Each solder ball has an average diameter of 1 mm and an average collapsed height of 0.5 mm. The package substrate uses a 1.27 mm thick *GETEK*© substrate with traces designed for an impedance of 50 Ω using trace widths of 0.1 mm. The wire bond connections to the package substrate are formed

Table 9.1 Electrical parasitic magnitudes for studied packages

	QFP-WB	BGA-WB	BGA-FC
L_{11}	4.550 nH	3.766 nH	1.244 nH
K_{12}	0.744	0.537	0.330
K_{13}	0.477	0.169	0.287
K_{14}	0.352	0.123	0.230
K_{15}	0.283	0.097	0.200
C_0	300 fF	288 fF	197 fF
C_{12}	121 fF	115 fF	96 fF
C_{13}	17 fF	97 fF	9 fF

Table 9.2 Electrical parasitics for various wire bond lengths

Length (mm)	C_{wb}(fF)	L_{wb}(nH)	$Z_0(\Omega)$
1	26	0.569	148
2	52	1.138	148
3	78	1.707	148
4	104	2.276	148
5	130	2.845	148

using 25 μm diameter gold wires to 100 × 400 μm wedge bond pads. The wire connections to the IC are formed using ball bonds onto 100 × 100 μm pads. The wire bonds range is length from 1 to 5 mm.

9.4.3 *Ball Grid Array with Flip-Chip Bumping*

The BGA-FC package studied in this work uses the same level 2 construction as the BGA-WB described above. The level 1 interconnect is formed using a 100 × 100 array of flip-chip bumps that have an average diameter of 125 μm and average collapsed height of 75 μm. The flip-chip bumps connect complimentary arrays of pads on the package substrate and on the IC. The pads on each substrate are 100 μm in diameter.

Table 9.1 lists the electrical parameter extraction results for the three packages described above. For each extraction the worst-case interconnect paths are extracted. These values represent the summation of all parasitic sources within the package.

The wire bond is the largest contributor of inductance in a bonded package. This interconnect is the largest source of impedance discontinuity and, in turn, the largest source of reflected noise. Chapter 15 presents an impedance compensation technique for this inductive interconnect so the specific electrical parasitics of a reasonable range of wire bond lengths were also extracted. Table 9.2 lists the specific inductance and capacitance parameters for a reasonable range of wire bond lengths used in the QFP-WB and BGA-WB packages.

Chapter 10
Preliminaries and Terminology

10.1 Bus Construction

A bus is defined as a group of signals that transmits data from one circuit to another. When designing a bus that traverses a package, the number of physical interconnect paths must be accounted for. This means that in addition to the total number of signal pins that are needed, the number of power and ground pins associated with the bus must also be considered. In VLSI design there are typically an equal number of power supply pins (V_{DD}) as ground pins (V_{SS}) [40, 59] for a given bus. For an arbitrarily large bus, individual bus *segments* can be defined that represent a subset of signals within a bus that are associated with at least one V_{DD} and at least one V_{SS} pin. By introducing the notation of a segment, subsets of the bus can be evaluated for performance instead of the entire bus, which leads to faster computation times without any loss of electrical accuracy.

Definition 10.1 A bus *segment* consists of a group of contiguous signal pins, along with at least one V_{DD} and at least one *GND* pin. The V_{DD} (*GND*) pins are also contiguous in their placement.

In practice, buses are implemented in the form of concatenated segments, with each segment having their V_{DD} and *GND* pins on either side of the signal pins.

Definition 10.2 Let $N_{segment}$ represent the total number of pins within the segment.

Definition 10.3 Let W_{bus} represent the number of signal pins within an individual segment.

Definition 10.4 Let N_g represent the number of V_{SS} pins within the bus segment where $N_g \geq 1$.

Definition 10.5 Let N_p represent the number of V_{DD} pins within the bus segment where $N_p \geq 1$.

This chapter describes the terminology and notation used throughout the rest of this monograph.

C. Duan et al., *On and Off-Chip Crosstalk Avoidance in VLSI Design*,
DOI 10.1007/978-1-4419-0947-3_10, © Springer Science+Business Media LLC 2010

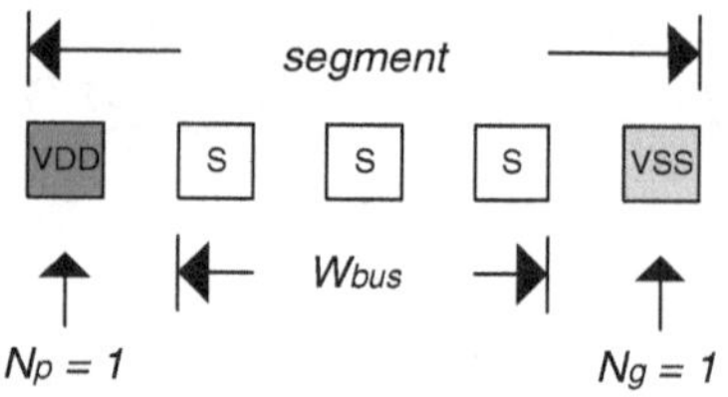

Fig. 10.1 Individual bus segment construction

Figure 10.1 shows an example segment. In this segment there are three signal pins ($W_{bus} = 3$), one V_{SS} pin ($N_g = 1$), and one V_{DD} pin ($N_p = 1$).

The ratio of signal-to-supply pins are defined to give a quick estimate of the quality of the power and ground distribution is for a given bus segment.

Definition 10.6 The *Signal-to-Power-Ground* (*SPG*) metric is defined as $W_{bus} : N_p : N_g$.

Definition 10.7 The *Signal-to-Power-Ground Ratio* (*SPGR*) is defined as W_{bus}/N_p. It assumes that $N_p = N_g = 1$.

In practice, most packages use $N_p = N_g = 1$. For the example segment in Figure 10.1, $SPG = 3 : 1 : 1$ and $SPGR = 3$.

The segment representation above corresponds to the commonly practiced bus construction in the VLSI industry today. We refer to segments by their index j, to keep trace of the relative location of a segment with respect to others.

Definition 10.8 Suppose a given bus has k segments. We represent an arbitrary segment of interest by its index j. Segments to the left of the jth segment are assigned indices $j-1, j-2, \ldots, j-(k/2)$. Segments to the right of the jth segment are assigned indices $j+1, j+2, \ldots, j+(k/2)$.

Definition 10.9 Let N_{bus} represent the total number of pin in an off-chip bus.

N_{bus} includes all signal, power, and ground pins that are used in the package to construct the off-chip bus. Figure 10.2 shows an example bus construction that consists of 3 segments ($k = 3$). In this bus each segment has the same number of signal ($W_{bus} = 3$), power ($N_p = 1$), and ground ($N_g = 1$) pins.

This notation allows a flexible framework that can represent arbitrarily large busses. Further, this notation matches the practical implementation of buses in VLSI

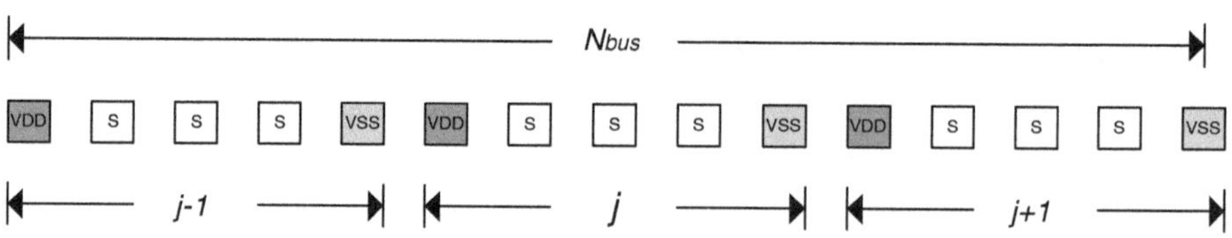

Fig. 10.2 Bus construction using multiple segments

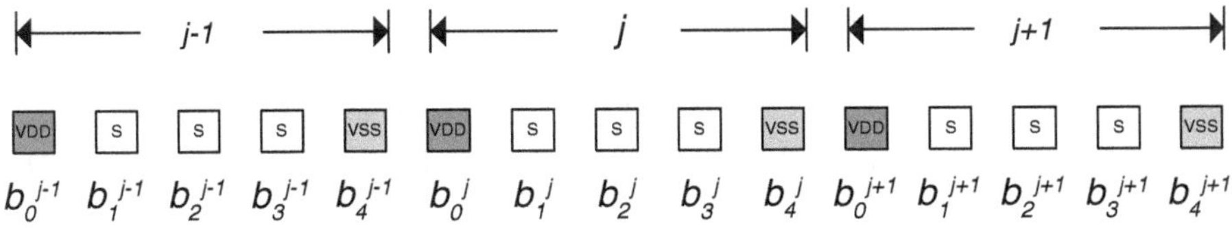

Fig. 10.3 Pin representation in a bus

designs. The number of pins within a segment is shown in Eq. 10.1.

$$N_{segment} = W_{bus} + N_p + N_g \tag{10.1}$$

The number of a pins within the entire bus is defined in Eq. 10.2.

$$N_{bus} = N_{segment} \cdot k \tag{10.2}$$

In order to specify the exact location within a bus and/or segment, an individual pin notation is defined. Each pin within a bus segment is given the notation b. Within a given segment, i is defined as the relative pin location.

Definition 10.10 The ith pin of the jth segment is denoted b_i^j. The indices i and j start with 0 and increase from left to right.

This framework allows the spatial location of any pin b_i^j within a bus to be described. Figure 10.3 shows an example bus using the spatial pin notation b_i^j. The arithmetic in the subscript of b_i^j is performed modulo $N_{segment}$. This allows a consistent method to describe relative locations of pins across segment boundaries. For example, if $N_{segment} = 5$, then $b_2^j = b_{2+5}^{j-1} = b_{2-5}^{j+1}$.

10.2 Logic Values and Transitions

Definition 10.11 s_i^j represents the logic state of pin b_i^j. For a signal pin, s_i^j can take on a value of 0 or 1 which represents the Boolean logic value for pin b_i^j. For a V_{SS} pin, $s_i^j = 0$, and for a V_{DD} pin, $s_i^j = 1$.

Definition 10.12 v_i^j represents the transition on any pin b_i^j. $v_i^j = 0$ when no transition occurs on pin b_i^j. $v_i^j = 1$ when pin b_i^j transitions from a logic 0 to a logic 1. $v_i^j = -1$ when pin b_i^j transitions from a logic 1 to a logic 0. $v_i^j = 0$ for V_{SS} or V_{DD} pins.

Figure 10.4 shows an example bus, illustrating the notation for the state and transition of each pin.

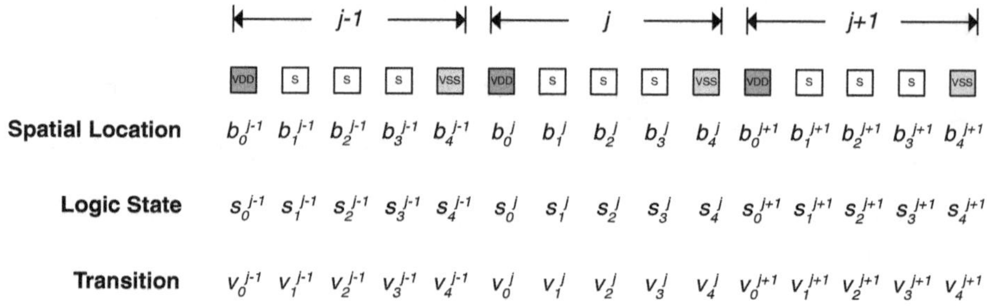

Fig. 10.4 STATE and transition notation

10.3 Signal Coupling

10.3.1 Mutual Inductive Signal Coupling

The between signals (M_{ik} or K_{ik}) represents the inductive coupling magnitude between an arbitrary pin b_i^j to its adjacent neighboring pins. The inductive coupling can span segment boundaries. Mutual inductive coupling exists between any two current carrying interconnects; however, when the magnitude of the coupling coefficient is relatively small the coupling can be ignored. The term p_L represents the number of neighbors on either side of pin b_i^j for which inductive coupling is considered. By considering only p_L neighbors on either side of the signal of interest, we simplify the analysis, without significantly compromising the quality of the results.

Figure 10.5 illustrates this idea. In this Figure, $p_L = 4$, which means that we consider 4 neighbors' mutual inductive coupling coefficients (to the left and to the right), and ignore the rest. Reducing p_L decreases the computation time when evaluating package noise, but results in larger error.

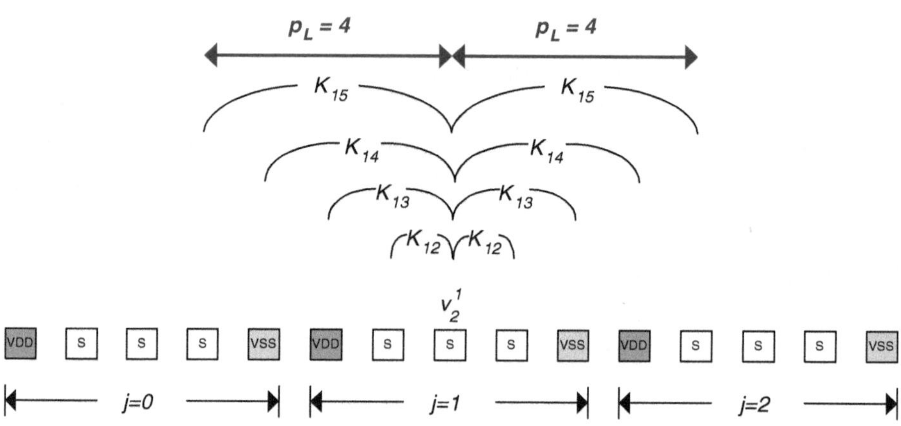

Fig. 10.5 Mutual inductive coupling notation

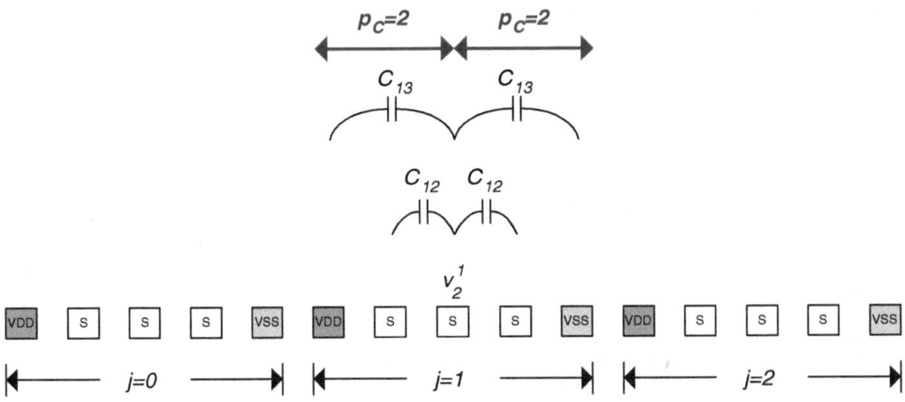

Fig. 10.6 Mutual capacitive coupling notation

10.3.2 Mutual Capacitive Signal Coupling

The between signals (C_{ik}) represents the capacitive coupling magnitude between an arbitrary pin b_i^j to its adjacent neighboring pins. In a similar manner to inductive coupling, mutual capacitive coupling can span segment boundaries. The term p_C is used to describe how many pins away from the pin of interest that capacitive coupling is considered. Typically, $p_C = 1$ is adequate for an accurate analysis, since mutual capacitances for subsequent neighbors are significantly smaller on account of capacitive shielding.

Figure 10.6 illustrates this idea. In this figure, $p_C = 2$, which means that we consider 2 neighbors' mutual capacitive coupling (to the left and right), and ignore the rest.

10.4 Return Current

When CMOS output drivers charge or discharge a signal line, will flow through the V_{SS} or V_{DD} pins in the package to complete the closed loop circuit path. When inductance is present in the V_{SS} or V_{DD} paths, then supply bounce will result. In VLSI packaging the return current for multiple signal pins will typically share a single supply pin. The return current will always seek the path of least resistance when traversing the package. In the case of VLSI packaging, this means that the return current will seek the supply or ground pin that is the spatially closest to the signal pin that is switching (due to the reduced inductance and resistance in the return path). This behavior may not appear to adhere to the *segment* notation described in Sect. 10.1. The actual return current for a pin may return current through another segment's supply pin; however, due to the symmetry of the segment notation, the total number of signals that utilize a segment's supply or ground pin is, in the worst-case, W_{bus}. This is due to the fact that even though signals within a particular segment may return

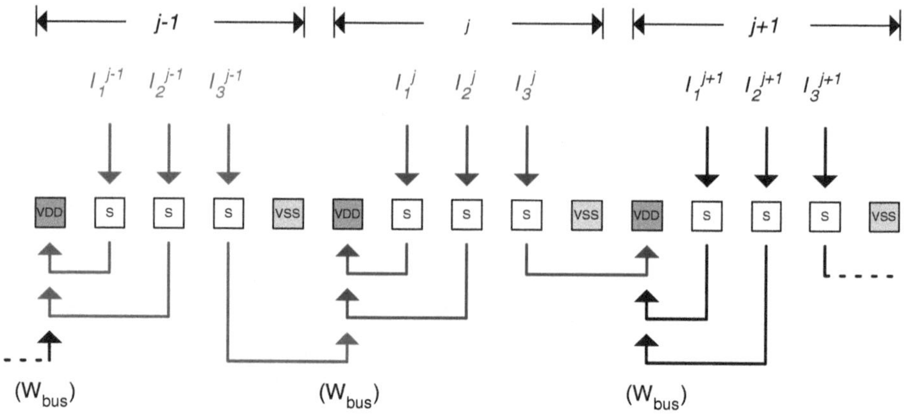

Fig. 10.7 Return current description

current through an adjacent segment's supply pin (which is spatially closer), signals from adjacent segments will return current through the supply pins of the segment of interest (because its supply pin is spatially closer to those signals). This allows the evaluation of an individual segment for supply bounce, considering only the signal pins within that segment. This does not lead to a reduction in modeling accuracy. This results in a much simpler analytical model for bus performance modeling and noise evaluation.

Figure 10.7 shows how the return current for a signal can span segment boundaries yet the total amount of return current within a segment will always be $W_{bus} \cdot I$. We assume that all the currents I_i^j are identical, and equal to I.

10.5 Noise Limits

The maximum allowable noise for any of the noise sources described in Sect. 8.2 is expressed in terms of user-defined parameters. For supply bounce, signal coupling, and reflected voltage, the noise limits are expressed as a percentage of V_{DD}. $P_x \cdot V_{DD}$ is the maximum allowable voltage noise defined by the user, where $0 \leq P_x \leq 1$. Figure 10.8 shows how the user-defined noise limits are expressed in terms of V_{DD}.

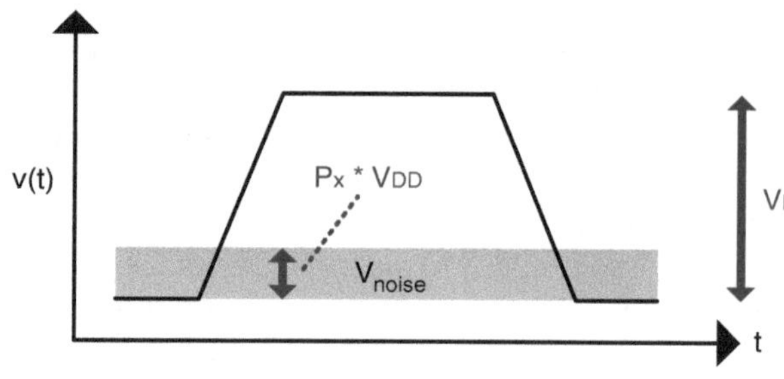

Fig. 10.8 Package noise notation

Definition 10.13 P_{gnd} represent the maximum allowable amount of ground bounce in the system. P_{gnd} is defined as a fraction of V_{DD} such that the noise due to ground bounce is $\leq P_{gnd} \cdot V_{DD}$.

Definition 10.14 N_{-1} represent the number of signal pins within a segment that are transitioning from a logic 1 to a logic 0. This number represents the number of signals that are returning current through the V_{SS} pin and resulting in ground bounce.

Definition 10.15 P_{supply} represent the maximum allowable amount of supply bounce in the system. P_{supply} is defined as a fraction of V_{DD} such that the noise due to supply bounce is $\leq P_{supply} \cdot V_{DD}$.

Definition 10.16 N_1 represent the number of signal pins within a segment that are transitioning from a logic 0 to a logic 1. This number represents the number of signals that are drawing current through the V_{DD} pin and resulting in supply bounce.

Definition 10.17 P_0 represent the maximum allowable amount of glitching noise in the system. P_0 is defined as a fraction of V_{DD} such that the absolute value of the glitching noise is $\leq P_0 \cdot V_{DD}$.

A glitch occurs when a victim signal is static while neighboring signals transition causing mutual inductive and capacitive coupling onto the victim line.

Definition 10.18 P_1 represent the maximum allowable amount of rising edge coupling noise in the system. P_1 is defined as a fraction of V_{DD} such that the rising edge coupling noise due to all neighboring signals is $\leq P_1 \cdot V_{DD}$.

Rising edge coupling occurs when a victim signal is transitioning from a logic 0 to a logic 1 at the same time that neighboring signals are transitioning and causing mutual inductive and capacitive coupling onto the victim line. This coupling can either speed up, slow down, or leave unaffected the rising edge on the victim line.

Definition 10.19 P_{-1} represent the maximum allowable amount of falling edge coupling noise in the system. P_{-1} is defined as a fraction of V_{DD} such that the falling edge coupling noise due to all neighboring signals is $\leq P_{-1} \cdot V_{DD}$.

Falling edge coupling occurs when a victim signal is transitioning from a logic 1 to a logic 0 at the same time that neighboring signals are transitioning and causing mutual inductive and capacitive coupling onto the victim line. This coupling can either speed up, slow down, or leave unaffected the falling edge on the victim line.

Definition 10.20 P_Γ represent the maximum allowable amount of reflected noise in the system. P_Γ is defined as a fraction of V_{DD} such that the absolute value of the reflected noise is $\leq P_\Gamma \cdot V_{DD}$.

When expressing the noise limit for capacitive bandwidth limiting, it is not applicable to use a fraction of V_{DD}; instead, the fraction by which the risetime of the signal is increased due to switching neighboring pins is used.

Definition 10.21 P_{BW} represent the maximum amount that the *original* risetime of a signal is altered due to switching neighboring pins. The *original* risetime is defined

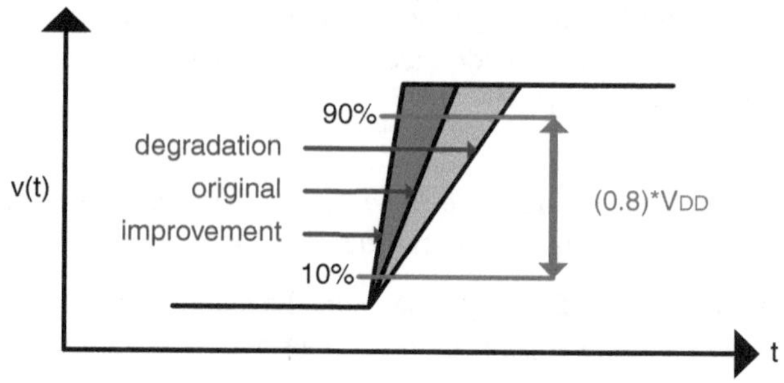

Fig. 10.9 Bandwidth limitation notation

as the time it takes to switch from 10% of V_{DD} to 90% of V_{DD} while all other signal pins are held at a logic 0.

When neighboring signal pins switch they change the effective capacitance that the victim signal experiences. This can cause the victim signal risetime to speed up, slow down, or remain the same. The fraction of risetime alteration is positive when the original risetime is sped up due to neighboring signal pins that switch. The fraction of risetime alteration is negative when the original risetime is slowed down due to neighboring signal pins that switch. Equation 10.3 gives the expression for how P_{BW} is defined. Figure 10.9 illustrates the mechanism of bandwidth limiting and how it relates to P_{BW}.

$$P_{BW} = \frac{t_{orig} - t_{new}}{t_{orig}} \tag{10.3}$$

Chapter 11
Analytical Model for Off-Chip Bus Performance

The noise sources described in Sect. 8.2 limit the maximum performance that an off-chip bus can achieve. This occurs because the package noise sources are proportional to the rate of change in current or voltage ($\frac{di}{dt}$ and $\frac{dv}{dt}$). When any given noise source exceeds the user-defined limits described in Sect. 10.5, the rate of change of current or voltage must be reduced.

11.1 Package Performance Metrics

The performance of an off-chip bus depends on how fast the digital output drivers can switch the output voltage levels. This relates to how much data that one pin of the bus can transmit per second. The term *Unit Interval* (UI) represents the shortest amount of time that data can be present on an individual pin and still accurately transmit the logic value from the Transmitter (Tx) to the Receiver (Rx) in the off-chip bus. The UI of a pin is directly related to the maximum switching speed that the system can achieve. Figure 11.1 shows a graphical representation of the Unit Interval metric.

The minimum can be translated into the maximum *Datarate* (DR) of an individual pin using Eq. 11.1.

$$DR_{max} = \frac{1}{UI_{min}} \tag{11.1}$$

Since the off-chip bus segment is constructed using multiple signal pins, the maximum *Throughput* (TP) of the bus segment can be found using Eq. 11.2. Figure 11.2 shows a graphical representation of throughput.

$$TP_{max} = DR_{max} \cdot W_{bus} \tag{11.2}$$

C. Duan et al., *On and Off-Chip Crosstalk Avoidance in VLSI Design*,
DOI 10.1007/978-1-4419-0947-3_11,

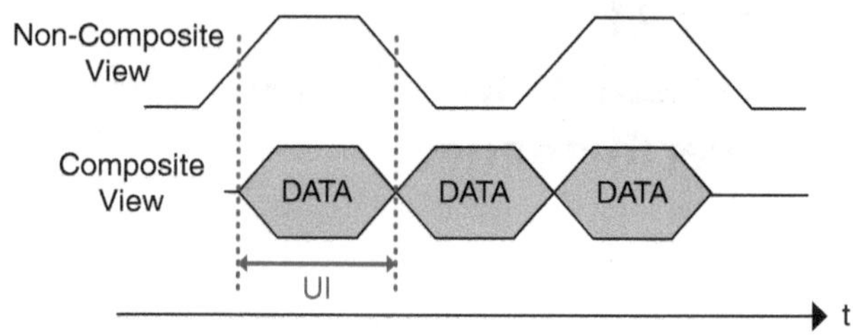

Fig. 11.1 Unit interval description

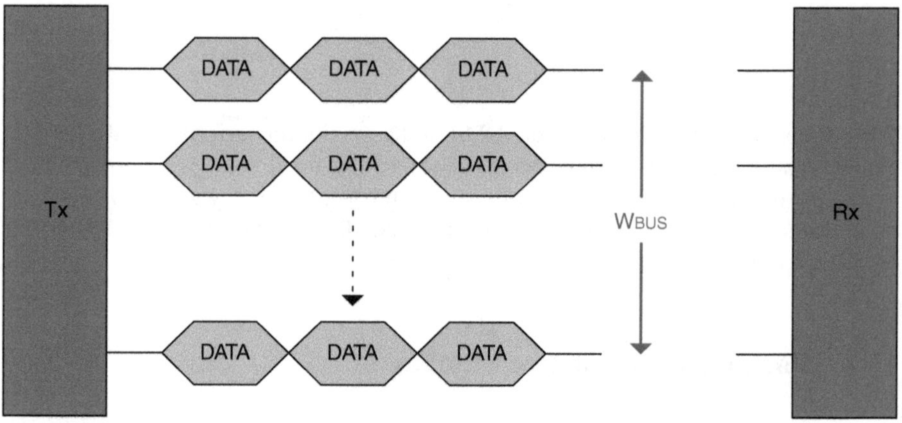

Fig. 11.2 Bus throughput description

11.2 Converting Performance to Risetime

To express the package performance metrics *UI*, *DR*, and *TP* in terms of rate of change of current or voltage, they must first be converted into the metric of *risetime*. The risetime (t_r) is defined as the time it takes to switch a signal from 10% of V_{DD} to 90% of V_{DD} (a total excursion of $0.8 \cdot V_{DD}$). Figure 11.3 shows the risetime definition.

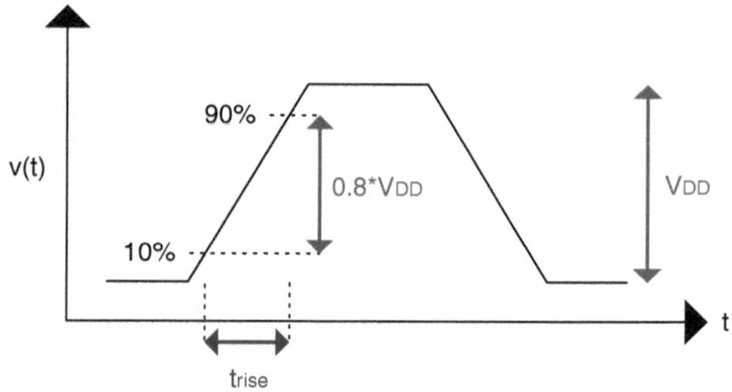

Fig. 11.3 Risetime description

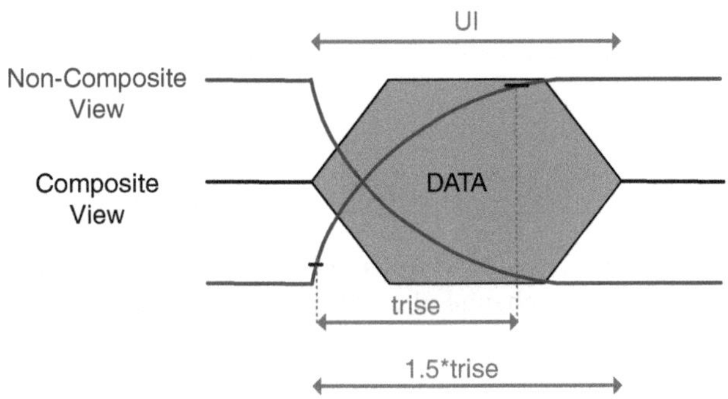

Fig. 11.4 Risetime to unit interval conversion

When a digital signal switches, the signal must be given adequate time to reach its steady state value before the next switching event occurs. The amount of time that is required is $UI_{\min}$ and can be expressed in terms of risetime (Eq. 11.3). For a robust digital system, the Unit Interval must be greater than or equal to 1.5 times the risetime [32, 58]. Figure 11.4 shows the graphical representation of how risetime relates to the minimum Unit Interval.

$$UI_{min} = 1.5 \cdot t_{rise} \tag{11.3}$$

11.3 Converting Bus Performance to $\frac{di}{dt}$ and $\frac{dv}{dt}$

Once the package performance has been expressed in terms of risetime, it can then be converted into the rate of change of current or voltage. The first step is to convert risetime into *slewrate*. Slewrate is defined as the rate of change in voltage ($\frac{dv}{dt}$) and typically is expressed in the units of $\frac{V}{ns}$. Slewrate can also be converted to the corresponding $\frac{di}{dt}$ by dividing it by the characteristic impedance of the system in which the package resides. Equation 11.4 expresses slewrate in terms of the rate of change in current or voltage. Figure 11.5 shows a graphical representation of slewrate.

$$slewrate = \frac{dv}{dt} = \frac{di}{dt} \cdot Z_0 \tag{11.4}$$

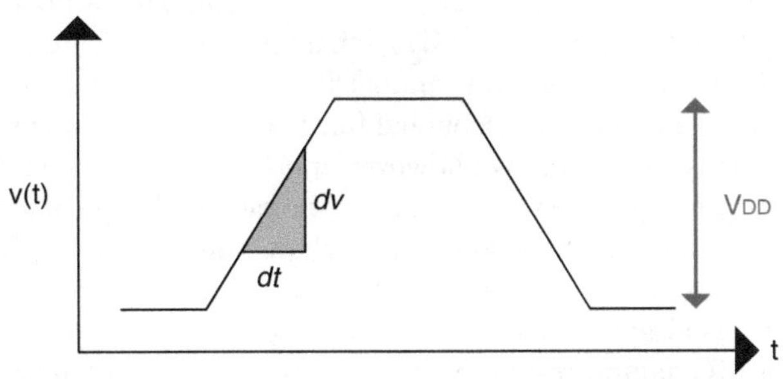

Fig. 11.5 Slewrate description

Using the definition of risetime as the time taken to switch from 10% of V_{DD} to 90% of V_{DD} (an excursion of $0.8 \cdot V_{DD}$), risetime can be expressed in terms of slewrate by Eq. 11.5.

$$t_{rise} = \frac{0.8 \cdot V_{DD}}{slewrate} \tag{11.5}$$

This, in turn, can be converted to a relationship between the package performance and the rate of change of current or voltage using Eqs. 11.6 and 11.7.

$$DR_{max} = \frac{\left(\frac{dv}{dt}\right)}{(1.5) \cdot (0.8) \cdot (V_{DD})} = \frac{\left(\frac{di}{dt}\right) \cdot Z_0}{(1.5) \cdot (0.8) \cdot (V_{DD})} \tag{11.6}$$

$$TP_{max} = \frac{\left(\frac{dv}{dt}\right) \cdot W_{bus}}{(1.5) \cdot (0.8) \cdot (V_{DD})} = \frac{\left(\frac{di}{dt}\right) \cdot Z_0 \cdot W_{bus}}{(1.5) \cdot (0.8) \cdot (V_{DD})} \tag{11.7}$$

11.4 Translating Noise Limits to Performance

Now that the noise sources described in Sect. 8.2 and the package performance metrics in Sect. 11.1 have been expressed in terms of the rate of change in current or voltage, a relationship between the noise limits and package performance can be constructed. For each noise source, the maximum datarate and throughput is found that does not violate any user-defined noise limit parameter. The noise source that approaches its user-defined limit first is the limiting source to performance.

11.4.1 Inductive Supply Bounce

When considering the performance limitation due to inductive supply bounce the fastest rate of $\frac{di}{dt}$ is found that results in the user-defined noise limits P_{supply} or P_{gnd} being violated. The first step is to find an expression for the total amount of supply bounce that results as a consequence of signal switching. The total amount of supply bounce is evaluated over one bus *segment*. The evaluation must include all signals within the bus segment (W_{bus}) that return current through the supply pin inductance (L_{11}). In addition, any mutual inductive and capacitive signal coupling onto the supply pin must be accounted for. Supply bounce and mutual inductive coupling are already in terms of $\frac{di}{dt}$; however, mutual capacitive coupling must first be converted into an expression that considers the rate of change of current or voltage. In order to do this, Eq. 8.9 is modified to include how the forcing function (t_{rise}) interacts with the natural response (τ_{RC}) of the capacitively coupled circuit [21, 32]. Performing the inverse Laplace Transform on an exponential ramp forcing function and assuming an RC natural response, the capacitive voltage divider is multiplied by (τ_{RC}/t_{rise})

to account for the risetime of the input voltage source. Equation 11.8 expresses the magnitude of the capacitively coupled voltage.

$$V_{vic} = \left(\frac{C_{1k}}{C_{1k} + C_0}\right) \cdot \Delta V_{agr} \cdot \left(\frac{\tau_{RC}}{t_{rise}}\right) \tag{11.8}$$

The time constant of the capacitively coupled circuit is given by Eq. 11.9.

$$\tau_{RC} = Z_0 \cdot (C_{1k} + C_0) \tag{11.9}$$

Using Eq. 11.5 to express t_{rise} in terms of $\frac{dv}{dt}$ (and $\frac{di}{dt}$) and substituting V_{DD} for ΔV_{agr}, the magnitude of the capacitively coupled signal can be expressed as in Eq. 11.10.

$$V_{vic} = \frac{C_{1k} \cdot Z_0 \cdot \left(\frac{dv}{dt}\right)}{(0.8)} = \frac{C_{1k} \cdot Z_0^2 \cdot \left(\frac{di}{dt}\right)}{(0.8)} \tag{11.10}$$

Equation 11.11[1] expresses the total amount of supply bounce in an off-chip segment. This expression represents the worst-case supply bounce pattern, which occurs when all signals within the segment switch from a logic 0 to a logic 1. This bus pattern means that all of the signal pins will charge their outputs from a 0 to a 1 by drawing current through the V_{DD} pin. In addition, this pattern causes mutual inductive and capacitive coupling to occur on the V_{DD} pin. This coupling adds to the total supply bounce noise in the system.

$$V_{Supply\text{-}Bnc} = \left(\frac{L_{11} \cdot W_{bus}}{N_p}\right) \cdot \left(\frac{di}{dt}\right) + \sum_{k=-p_L}^{p_L} M_{1(|k|+1)} \cdot \left(\frac{di}{dt}\right) + \sum_{k=-p_C}^{p_C} \frac{C_{1(|k|+1)} \cdot Z_0^2}{(0.8)} \cdot \left(\frac{di}{dt}\right) \tag{11.11}$$

We set Equation 11.11 to the user-defined maximum allowed supply bounce (which is expressed as a fraction P_{supply} of V_{DD}), to get Eq. 11.12.

$$P_{supply} \cdot V_{DD} = \left(\frac{L_{11} \cdot W_{bus}}{N_p}\right) \cdot \left(\frac{di}{dt}\right) + \sum_{k=-p_L}^{p_L} M_{1(|k|+1)} \cdot \left(\frac{di}{dt}\right) + \sum_{k=-p_C}^{p_C} \frac{C_{1(|k|+1)} \cdot Z_0^2}{(0.8)} \cdot \left(\frac{di}{dt}\right) \tag{11.12}$$

[1]Note in this equation that when the subscript evaluates to M_{11} or C_{11}, these values are set equal to zero.

By factoring $\frac{di}{dt}$ out of Eq. 11.12 and substituting into Eq. 11.6, the maximum datarate that can be achieved without violating the user-defined noise limit for supply bounce can be found (Eq. 11.13).

$$DR_{max-supply} = \frac{P_{supply} \cdot Z_0}{(1.5) \cdot (0.8) \cdot \left[\left(\frac{L_{11} \cdot W_{bus}}{N_p} \right) + \sum_{-PL}^{PL} M_{1(|k|+1)} + \sum_{-PC}^{PC} \frac{C_{1(|k|+1)} \cdot Z_0^2}{(0.8)} \right]} \quad (11.13)$$

This, in turn, can be translated into the maximum throughput of the segment by multiplying it by the number of signals within the segment (Eq. 11.14).

$$TP_{max-supply} = \frac{P_{supply} \cdot Z_0 \cdot W_{bus}}{(1.5) \cdot (0.8) \cdot \left[\left(\frac{L_{11} \cdot W_{bus}}{N_p} \right) + \sum_{-PL}^{PL} M_{1(|k|+1)} + \sum_{-PC}^{PC} \frac{C_{1(|k|+1)} \cdot Z_0^2}{(0.8)} \right]} \quad (11.14)$$

In a similar manner, the maximum datarate and throughput for a system (such that the user-defined noise limit P_{gnd} is not violated) can be derived. These expressions are given in Eqs. 11.15 and 11.16.

$$DR_{max-gnd} = \frac{P_{gnd} \cdot Z_0}{(1.5) \cdot (0.8) \cdot \left[\left(\frac{L_{11} \cdot W_{bus}}{N_g} \right) + \sum_{-PL}^{PL} M_{1(|k|+1)} + \sum_{-PC}^{PC} \frac{C_{1(|k|+1)} \cdot Z_0^2}{(0.8)} \right]} \quad (11.15)$$

$$TP_{max-gnd} = \frac{P_{gnd} \cdot Z_0 \cdot W_{bus}}{(1.5) \cdot (0.8) \cdot \left[\left(\frac{L_{11} \cdot W_{bus}}{N_g} \right) + \sum_{-PL}^{PL} M_{1(|k|+1)} + \sum_{-PC}^{PC} \frac{C_{1(|k|+1)} \cdot Z_0^2}{(0.8)} \right]} \quad (11.16)$$

11.4.2 Capacitive Bandwidth Limiting

When considering the in the package, the slowest RC time constant is found which corresponds to the worst-case switching pattern on the bus. This time constant is then translated into risetime, which in turn is translated into datarate and throughput as in the previous section. For capacitive bandwidth limitation, the worst-case pattern on the bus occurs when the center pin within the segment transitions from a logic 0 to a 1 while all other signals switch from a 1 to a 0 (or vice versa). This switching pattern results in the highest effective capacitance on the victim signal. This capacitance will result in the slowest risetime since it results in the largest RC time constant for the bus segment. This dictates the maximum datarate and throughput that can be achieved in the package. Combining Eqs. 8.7 and 8.8 gives the maximum risetime that a given package can achieve when considering capacitive bandwidth limitation.

$$t_{rise-BW} = 2.2 \cdot Z_0 \cdot \left(C_0 + \sum_{k=-pc}^{pc} 2 \cdot C_{1(|k|+1)} \right) \tag{11.17}$$

This can be directly converted to datarate and throughput using Eq. 11.3. Equations 11.18 and 11.19 give the expressions for the maximum datarate and throughput that a package can achieve when considering bandwidth limitation.

$$DR_{max-BW} = \frac{1}{(1.5) \cdot (2.2) \cdot Z_0 \cdot (C_0 + \sum_{-pc}^{pc} 2 \cdot C_{1(|k|+1)})} \tag{11.18}$$

$$TP_{max-BW} = \frac{W_{bus}}{(1.5) \cdot (2.2) \cdot Z_0 \cdot (C_0 + \sum_{-pc}^{pc} 2 \cdot C_{1(|k|+1)})} \tag{11.19}$$

11.4.3 Signal Coupling

The derivation of the magnitude is performed exactly as in Sect. 11.4.1. Equation 11.20 expresses the maximum amount of noise voltage due to mutually inductive and capacitive signal coupling onto a victim line. This expression describes the coupled noise when the victim signal is either static (P_0), rising (P_1), or falling (P_{-1}).

$$V_{coupling} = \sum_{k=-p_L}^{p_L} M_{1(|k|+1)} \cdot \left(\frac{di}{dt}\right) + \sum_{k=-pc}^{pc} \frac{C_{1(|k|+1)} \cdot Z_0^2}{(0.8)} \cdot \left(\frac{di}{dt}\right) \tag{11.20}$$

Equating this to the user-defined maximum allowed coupled noise (which is expressed as a fraction P_0, P_1, or P_{-1} of V_{DD}), we get Eq. 11.21.

$$P_{(0/-1/1)} \cdot V_{DD} = \sum_{k=-p_L}^{p_L} M_{1(|k|+1)} \cdot \left(\frac{di}{dt}\right) + \sum_{k=-pc}^{pc} \frac{C_{1(|k|+1)} \cdot Z_0^2}{(0.8)} \cdot \left(\frac{di}{dt}\right) \tag{11.21}$$

By factoring $\frac{di}{dt}$ out of Eq. 11.21 and substituting into Eq. 11.6, the maximum datarate that can be achieved without violating the user-defined noise limits for signal coupling can be found (Eq. 11.22).

$$DR_{max-coupling} = \frac{P_{(0/-1/1)} \cdot Z_0}{(1.5) \cdot (0.8) \cdot \left[\sum_{-p_L}^{p_L} M_{1(|k|+1)} + \sum_{-pc}^{pc} \frac{C_{1(|k|+1)} \cdot Z_0^2}{(0.8)} \right]} \tag{11.22}$$

Equation 11.23 expresses the maximum throughput when considering the user-defined noise limits for signal coupling.

$$TP_{max-coupling} = \frac{P_{(0/-1/1)} \cdot Z_0 \cdot W_{bus}}{(1.5) \cdot (0.8) \cdot \left[\sum_{-PL}^{PL} M_{1(|k|+1)} + \sum_{-PC}^{PC} \frac{C_{1(|k|+1)} \cdot Z_0^2}{(0.8)} \right]} \tag{11.23}$$

11.4.4 Impedance Discontinuities

When evaluating a package for transmission line reflections due to, the forcing function and natural response of the circuit must be considered in the evaluation of Eqs. 8.13 and 8.14. To accomplish this we assume that the natural response of the package behaves as a low-pass RC filter. This allows the same derivations made in Sect. 11.4.1 to be applied to the evaluation of the transmission line reflections. Equation 11.24 expresses the magnitude of the reflection, considering the forcing function and the natural response of the package.

$$V_{reflected} = V_{incident} \cdot (\Gamma) \cdot \left(\frac{\tau_{RC}}{t_{rise}} \right) \tag{11.24}$$

Substituting the RC time constant given in Equation 11.9 and replacing $V_{incident}$ with V_{DD}, the reflected voltage can be expressed in terms of the user-defined noise limit (which is expressed as a fraction P_Γ of V_{DD}).

$$P_\Gamma \cdot V_{DD} = V_{DD} \cdot (\Gamma) \cdot \left(\frac{Z_0 \cdot \left(\sum_{-PC}^{PC} C_{1(|k|+1)} + C_0 \right)}{t_{rise}} \right) \tag{11.25}$$

This can be substituted into Eq. 11.6 to find the maximum datarate that does not violate the user-defined noise limit P_Γ. Equations 11.26 and 11.27 give the datarate and throughput when considering the maximum allowable noise due to impedance discontinuities.

$$DR_{max-\Gamma} = \frac{P_\Gamma}{(1.5) \cdot \Gamma \cdot Z_0 \cdot \left(\sum_{-PC}^{PC} C_{1(|k|+1)} + C_0 \right)} \tag{11.26}$$

$$TP_{max-\Gamma} = \frac{P_\Gamma \cdot W_{bus}}{(1.5) \cdot \Gamma \cdot Z_0 \cdot \left(\sum_{-PC}^{PC} C_{1(|k|+1)} + C_0 \right)} \tag{11.27}$$

11.5 Experimental Results

In order to verify the analytical model, SPICE [76] simulations were performed on a test circuit. The test circuit includes a model for the package parasitics described in Table 9.1. The test circuit is used to monitor noise limit violations due to the package as the performance of the output driver is increased. By monitoring noise limit violations the simulation can indicate the maximum datarate and throughput for a given package and bus configuration. These simulation results are compared to the analytical model predictions to verify the model's accuracy.

11.5.1 Test Circuit

The test circuit used in this analysis consists of a standard CMOS inverter as the output driver. The CMOS inverter is implemented using the BPTM 0.1 μm [1] technology using BSIM3 model cards [2]. The CMOS inverter is sized to have an equal drive strength when outputting a logic 0 or a logic 1 by sizing the PMOS and NMOS transistors such that $(\frac{W_P}{W_N}) = (\frac{u_n}{u_p}) = 3.25$ [59]. The CMOS inverter drives through the Tx package model onto a PCB. The PCB contains a series termination resistor (R_s) followed by 2″ of 50 Ω transmission line. The receiver load is modeled using the input to a CMOS inverter that is sized to have a gate capacitance of $3\,pF$. A package model is also inserted in the signal and power path prior to the Rx inverter. Both the Tx and Rx inverters have $V_{DD} = 1.5$ v and $V_{SS} = 0$ v. The output risetime of the transmitter is altered by resizing the CMOS transistors in the inverter. Increasing the size of the transistor will result in a smaller risetime. Figure 11.6 shows the test circuit topology.

For each of the three packages studied in this work (QFP-WB, BGA-WB, and BGA-FC), the number of switching signal pins within a segment are varied from $W_{bus} = 1$ to $W_{bus} = 16$ in the simulation. In addition, three different SPG configurations are used to observe the effect of adding power and ground pins to the segment. The three SPG configurations used are SPG = 8 : 1 : 1, SPG = 4 : 1 : 1, and SPG = 2 : 1 : 1. The user-defined noise limits are set to 5% of V_{DD} so that

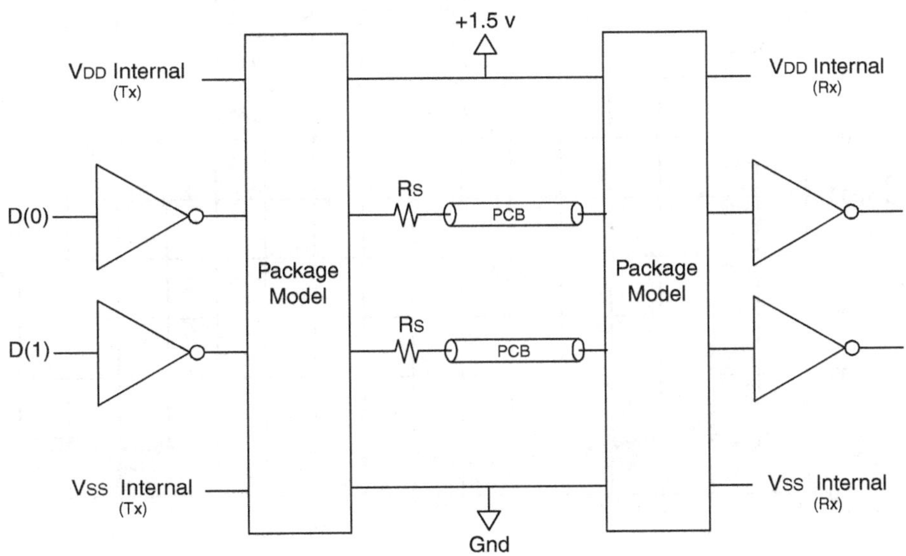

Fig. 11.6 Test circuit used to verify analytical model

$P_{supply} = P_{gnd} = P_0 = P_1 = P_{-1} = -P_{BW} = P_{\Gamma} = 0.05$. It was found that for the three packages studied, supply bounce (P_{supply}, P_{gnd}) was always the limiting factor to performance.

11.5.2 Quad Flat Pack with Wire Bonding Results

Figure 11.7 shows the per-pin datarate for the QFP-WB package as a function of the number of signal pins within the segment. This plot shows the predicted datarate for the analytical model versus the simulated results. These results illustrate that the per-pin datarate needs to be reduced as channels are added to the bus segment in order to keep the noise below the user-defined limits. This plot also shows the amount of ground and power pins per segment influences the performance of the segment. Segments with a lower SPR perform at higher datarates due to the reduction in the total $\frac{di}{dt}$ that is drawn through each supply pin.

Figure 11.8 shows the bus throughput for the QFP-WB package as a function of the number of signals pins within the segment. Due to the reduction in per-pin datarate as channels are added to the segment, throughput does not increase linearly with the addition of channels; instead, the throughput follows a less than linear increase as pins are added to the segment.

Figure 11.9 shows the error percentage between the simulation results and analytical model prediction for each size of segment evaluated in the QFP-WB package.

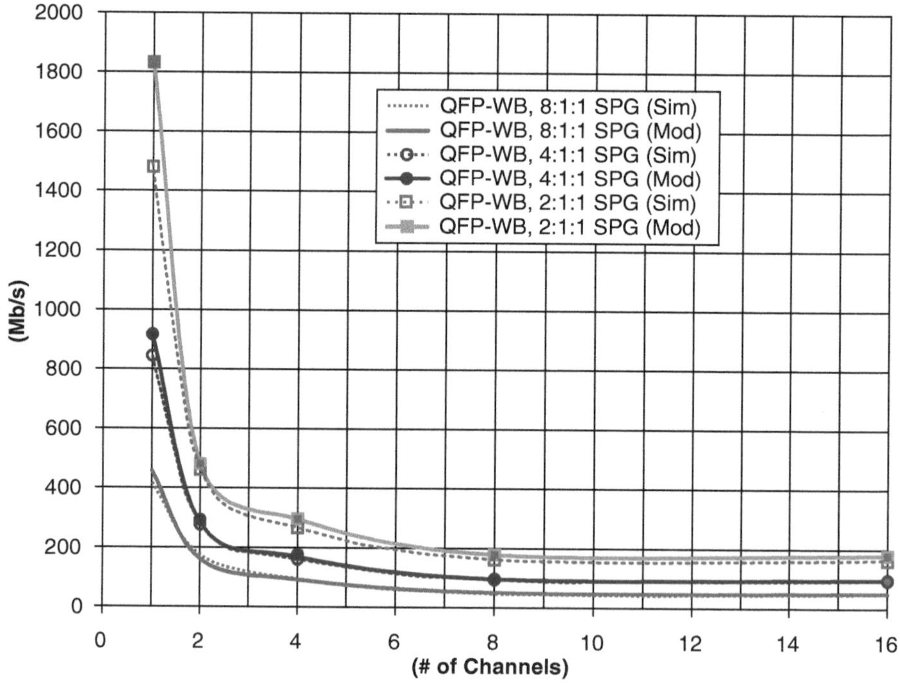

Fig. 11.7 Per-Pin datarate for a QFP-WB package

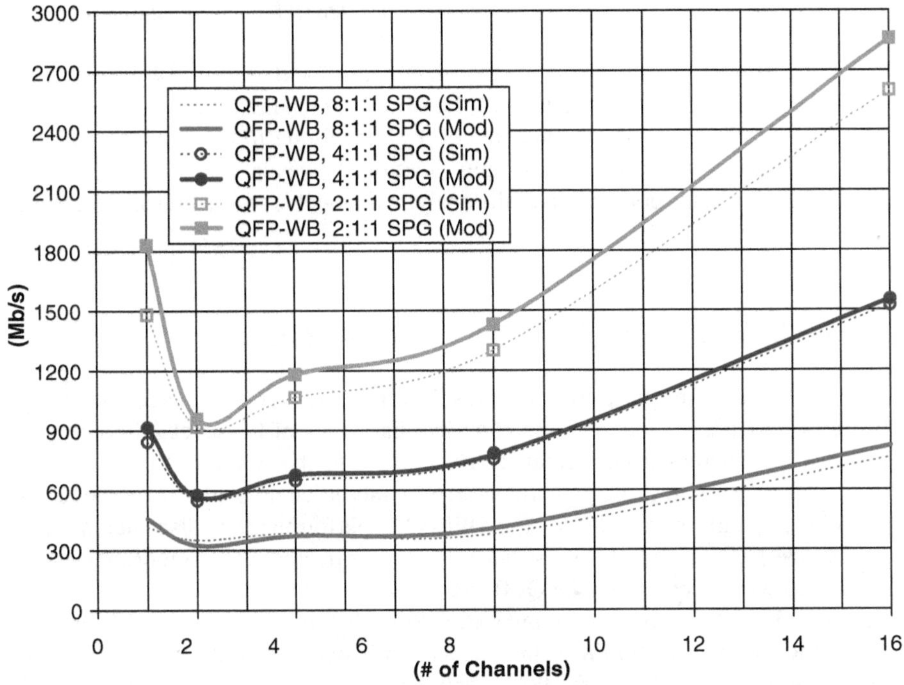

Fig. 11.8 Throughput for a QFP-WB package

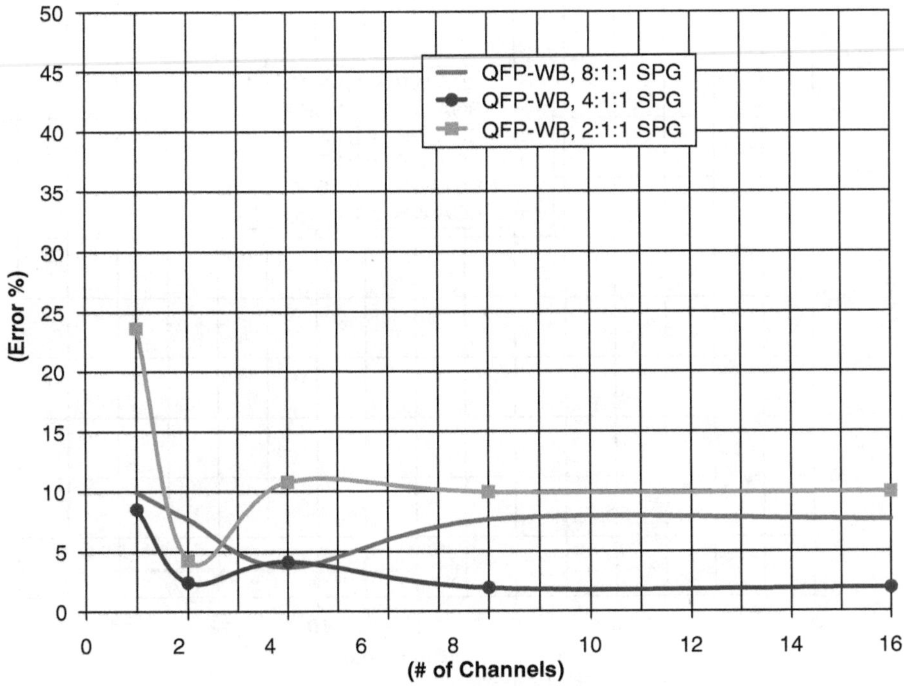

Fig. 11.9 Model accuracy for a QFP-WB package

This plot illustrates the accuracy of the analytical model. In almost all cases the accuracy of the model is within 10% of the simulated results. The analytical model exhibits the largest error for a bus size of 1 channel due to the higher frequency components being present in the forcing function; however, even with an error of 23% for 1 channel, the correlation of the analytical model is still within an acceptable range.

11.5.3 Ball Grid Array with Wire Bonding Results

Figure 11.10 shows the per-pin datarate results for the BGA-WB package. This figure illustrates how the reduction in the electrical parasitics of the level 2 interconnect translates into increased performance. By moving to the BGA-WB package with less electrical parasitics, the per-pin datarate is increased as much as 25% over the QFP-WB. As with the results for the QFP-WB, performance is also increased by using a lower SPG Ratio. Also, the per-pin datarate decreases with more switching channels as was observed for the QFP-WB package, for the same reasons.

Figure 11.11 shows the throughput for the BGA-WB package. Again, the throughput experiences a less than linear increase in performance as channels are added to the bus segment. In addition, the throughput suffers a plateau at low channel counts

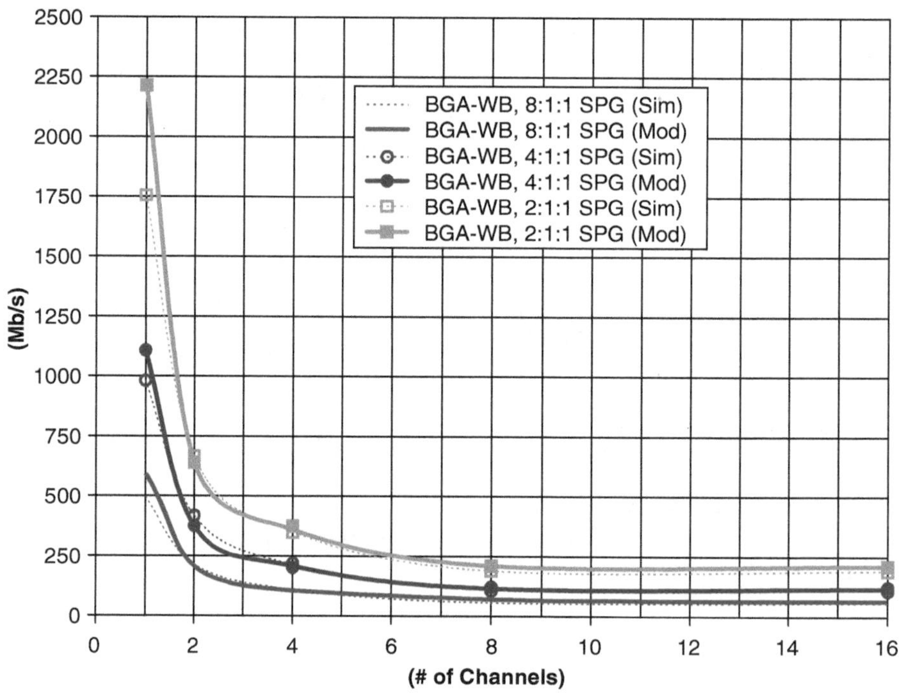

Fig. 11.10 Per-Pin datarate for a BGA-WB package

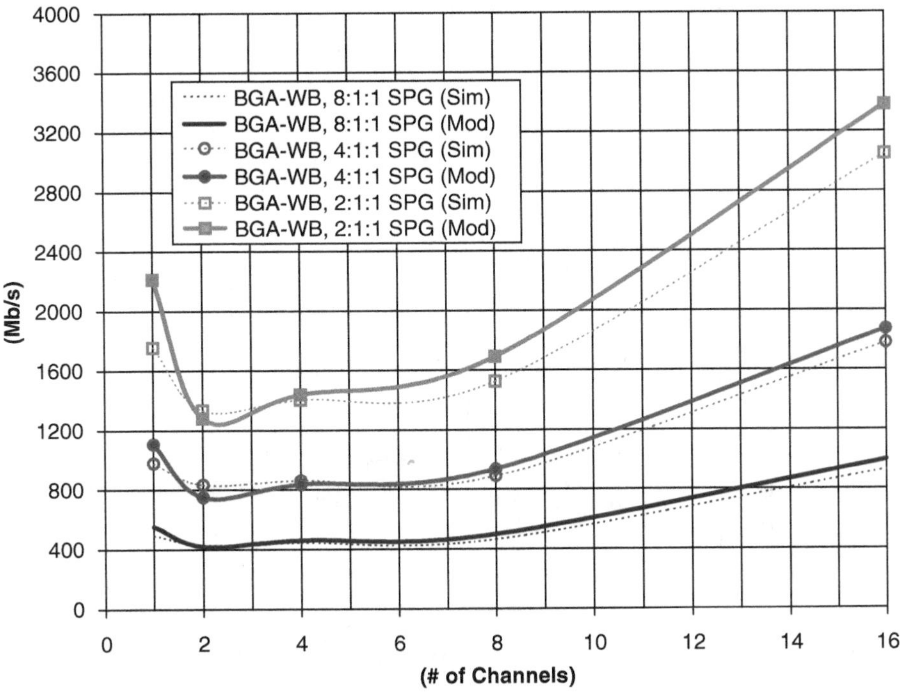

Fig. 11.11 Throughput for a BGA-WB package

at which the throughput is not increased as channels are added. This behavior leads to multiple bus configurations having the same throughput. In these cases the narrowest bus configuration can be chosen to reduce cost.

Figure 11.12 shows the amount of error between the analytical model prediction and the simulated results for the BGA-WB package. Again, the most error occurs for segments containing only one signal pin. For channel counts greater than 1, the model accuracy is consistently below 11%.

11.5.4 Ball Grid Array with Flip-Chip Bumping Results

Figure 11.13 shows the per-pin datarate results for the BGA-FC package. The BGA-FC package is the most advanced package studied in this work. By implementing flip-chip bumping in the level 1 interconnect in addition to a BGA in the level 2 interconnect, this package is able to achieve much higher performance over the QFP-WB and BGA-WB. The per-pin datarate achieves a 260% increase in performance over the BGA-WB package and a 333% increase over the QFP-WB. Despite its much higher performance, the BGA-FC still suffers a reduction in per-pin datarate as signals are added to the segment. A lower SPG also results in a higher per-pin datarate, as with the other packages studied.

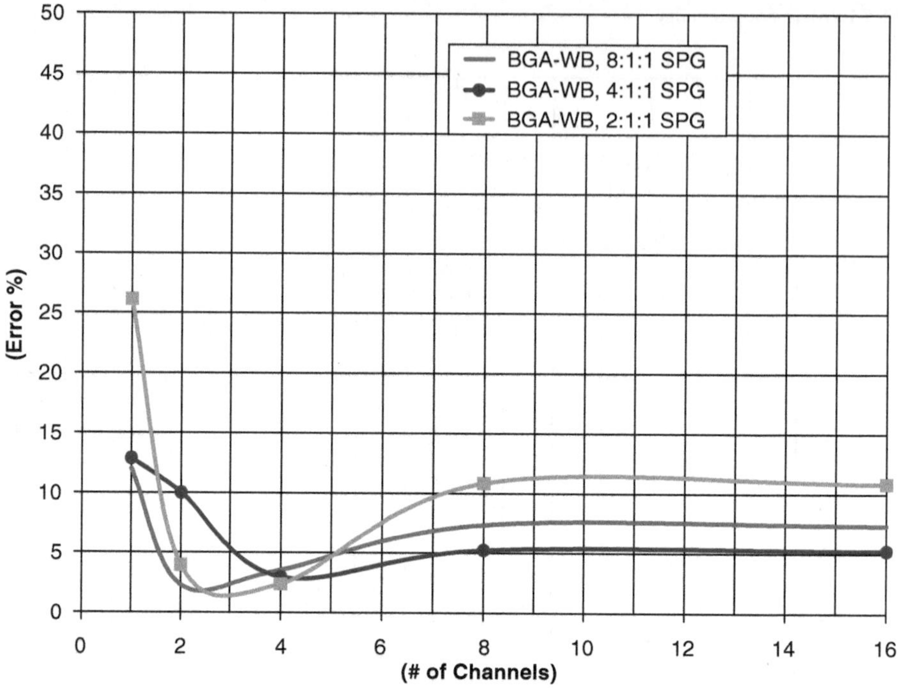

Fig. 11.12 Model accuracy for a BGA-WB package

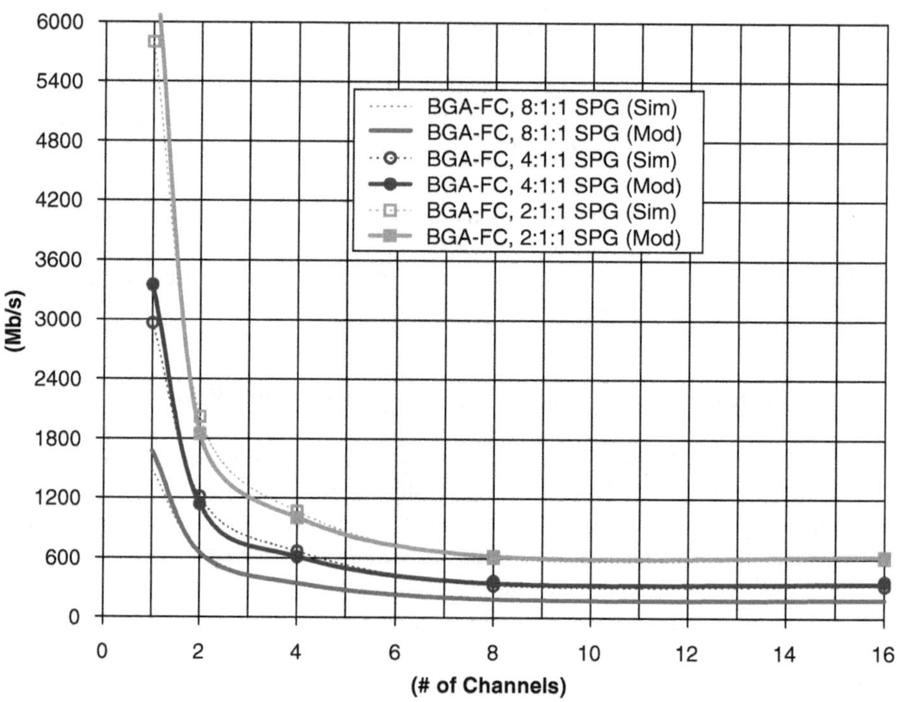

Fig. 11.13 Per-Pin datarate for a BGA-FC package

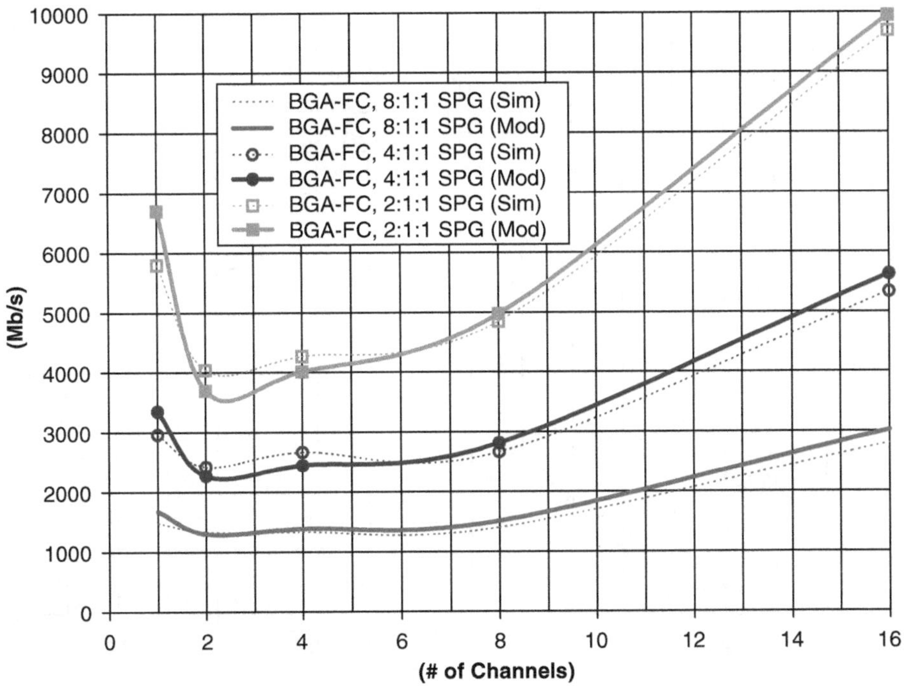

Fig. 11.14 Throughput for a BGA-FC package

Figure 11.14 shows the throughput for the BGA-FC package. Again, the throughput increases at a less than linear rate as channels are added to the segment. In addition, the BGA-FC package also exhibits the phenomenon where multiple bus configurations achieve very similar throughputs.

Figure 11.15 shows the accuracy of the analytical model for the BGA-FC package. The worst-case model error occurs at a segment size of 1 while all other configurations achieve less than a 10% error.

11.5.5 Discussion

Since the framework of this model is constructed in a parameterized and scalable fashion, it can be applied to differential as well as single ended signaling with minimal alteration. In differential signaling, a portion of the return current for a given signal pin will flow through its complementary pin within the differential pair. In this special case, the amount of supply bounce noise due to current being drawn through the inductance of the V_{DD} or V_{SS} pin is reduced by the amount of current that is inherently returned within each differential pair of the segment. The advantage of differential signaling is that since a portion of the return current is provided within the pair itself, this current does not flow through the inductance of the supply pin and does not result in supply bounce noise. All of the other noise sources that are predicted within this model do not require modification when using differential signaling.

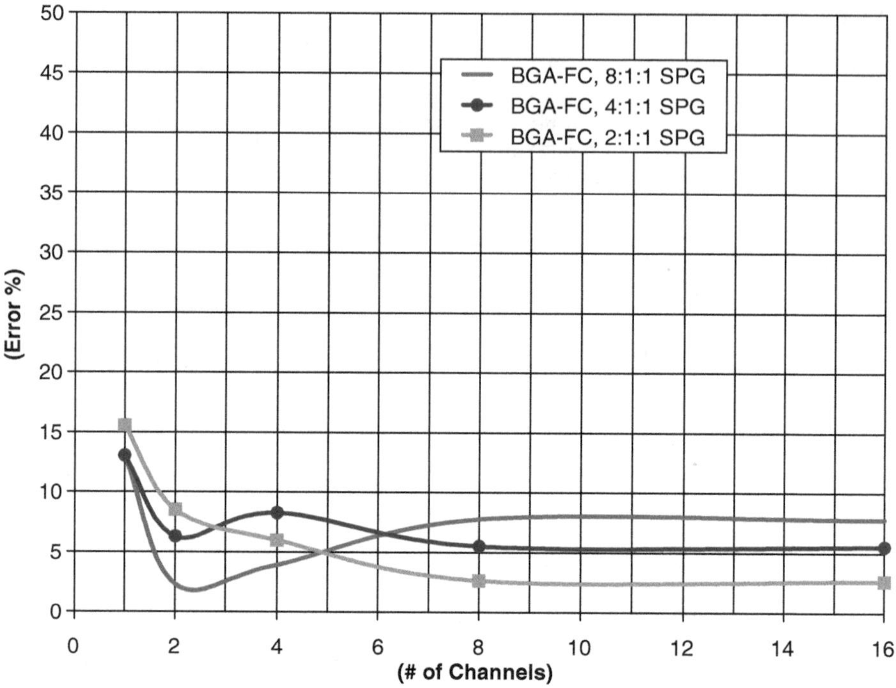

Fig. 11.15 Model accuracy for a BGA-FC package

These experimental results illustrate how the electrical parasitics of the IC package limit the overall system performance. The effect of increasing the number of simultaneously switching signals is that the per-pin performance is significantly reduced. In most cases a desired bus throughput can be achieved using multiple package and pin configurations, including different signal counts within a given segment. For these cases we must factor in the cost of each package and pin configuration in order to select the most economical bus design for a required throughput. In the next chapter, a methodology is provided to select the most cost-effective bus configuration for a given throughput requirement.

Chapter 12
Optimal Bus Sizing

The package performance results reported in Chap. 11 revealed that simply adding signals to a bus does not necessarily increase the throughput of the system. As more channels are added to the bus, the noise caused by simultaneously switching signals reduces the per-pin datarate. This behavior leads to a less than linear increase in system performance as channels are added. In some cases, the system performance actually decreases as more channels are added. In practice, multiple packages, configurations, and segment sizes are able to achieve the same throughput. In this chapter, a method is described for choosing the package, bus size, and SPG in the most cost-effective manner for a given throughput requirement.

12.1 Package Cost

The use of advanced packaging will reduce the electrical parasitics due to the package. This reduces the package noise in the system which leads to increased performance; however, moving toward advanced packaging is often more expensive and can result in a cost prohibitive increase in the majority of modern VLSI designs.

Table 12.1 lists the average cost per-pin ($Cost_{per\text{-}pin}$) for the three packages that have been studied in this work [40]. The QFP-WB package has the least cost per pin due to its mature manufacturing process and scalable interconnect; however, this package suffers the worst performance due to the relatively large inductance in the lead frame and wire bond interconnect. The BGA-WB improves upon the lead frame connection by moving toward a ball grid array. The BGA interconnect reduces the inductance in the level 2 interconnect (Table 9.1) but increases the per pin cost over the QFP-WB by 55%. The BGA-FC package experiences the best electrical performance by using flip-chip bumping for its level 1 interconnect (instead of wire bonding). The inductance in the BGA-FC package is significantly reduced over both the QFP-WB and BGA-WB as shown in Chap. 9; however, the improvement in performance comes at an increase in cost of 85% over the BGA-WB and 286% over the QFP-WB.

Table 12.2 lists the total number of pins that are needed to implement the various SPG configurations used in this work. The number of pins needed to implement a

C. Duan et al., *On and Off-Chip Crosstalk Avoidance in VLSI Design*,
DOI 10.1007/978-1-4419-0947-3_12,

Table 12.1 Package I/O cost (US Dollars, $)

Package	$Cost_{per\text{–}pin}$
QFP-WB	$0.22
BGA-WB	$0.34
BGA-FC	$0.63

Table 12.2 Number of pins needed per bus configuration

	Number of channels				
Bus configuration	1	2	4	8	16
QFP-WB 8:1:1	3	4	6	10	20
QFP-WB 4:1:1	3	4	6	12	24
QFP-WB 2:1:1	3	4	8	16	32
BGA-WB 8:1:1	3	4	6	10	20
BGA-WB 4:1:1	3	4	6	12	24
BGA-WB 2:1:1	3	4	8	16	32
BGA-FC 8:1:1	3	4	6	10	20
BGA-FC 4:1:1	3	4	6	12	24
BGA-FC 2:1:1	3	4	8	16	32

bus increases at a faster-than-linear rate as signals are added due to the need for a pair of V_{DD} and V_{SS} within each segment.

The total cost of the bus configuration is given by Eq. 12.1 where the cost is simply the per-pin cost multiplied by the number of pins needed to implement the bus segment. Table 12.3 shows the cost of the various bus configurations. The effect of choosing a better grounding and power scheme (i.e., a lower SPGR) is that the cost increases at a faster-than-linear rate as channels are added. This table shows the relative expense of the different packages considered and illustrates how moving toward advanced packaging can lead to a significant cost increase.

$$Cost_{bus} = (N_{bus}) \cdot (Cost_{per\text{–}pin}) \tag{12.1}$$

Table 12.3 Total cost for various bus configurations ($)

	Number of channels				
Bus configuration	1	2	4	8	16
QFP-WB 8:1:1	0.66	0.88	1.32	2.20	4.40
QFP-WB 4:1:1	0.66	0.88	1.32	2.62	5.28
QFP-WB 2:1:1	0.66	0.88	1.76	3.52	7.04
BGA-WB 8:1:1	1.02	1.36	2.04	3.40	6.80
BGA-WB 4:1:1	1.02	1.36	2.04	4.08	8.16
BGA-WB 2:1:1	1.02	1.36	2.72	5.44	10.88
BGA-FC 8:1:1	1.89	2.52	3.78	6.30	12.60
BGA-FC 4:1:1	1.89	2.52	3.78	7.56	15.12
BGA-FC 2:1:1	1.89	2.52	5.04	10.08	20.16

12.2 Bandwidth Per Cost

In order to compare the cost-effectiveness of a package configuration, the metric *Bandwidth-per-Cost* (*BPC*) is introduced. This metric has units of ($\frac{Mb}{\$}$) and takes into account the total bus throughput as well as the cost to implement a given bus configuration. Equation 12.2 gives the definition of the BPC metric. When comparing two busses the configurations, the configuration with a higher BPC indicates that it is able to transmit more data for less cost.

$$BPC = \left(\frac{TP}{Cost_{bus}}\right) \tag{12.2}$$

12.2.1 Results for Quad Flat Pack with Wire Bonding

Figure 12.1 shows the *Bandwidth-per-Cost* for the QFP-WB package as signals are added to the segment. This figure illustrates that it is more cost-effective to construct a bus that is narrow (and fast) rather than the traditional wider (and slower) configuration. These results match closely with the current industrial trends for high-end computer busses. Busses such as HyperTransport [30] and PCI Express [77] have taken advantage of the fact that narrower busses avoid the noise problems due to simultaneously switching signals, thus enabling higher per-pin datarates. At the same

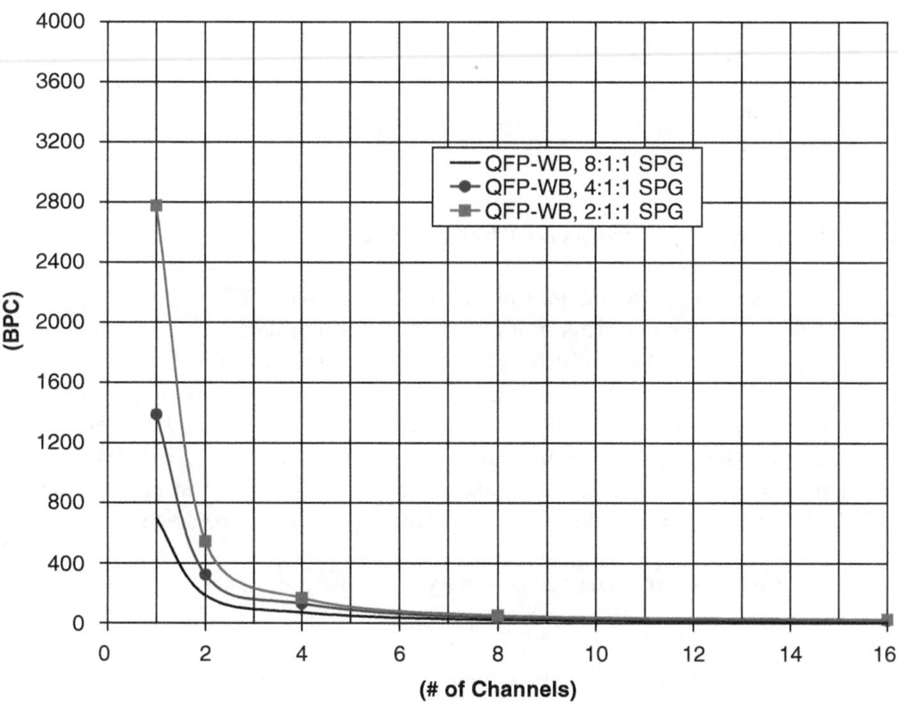

Fig. 12.1 Bandwidth-per-Cost for a QFP-WB package (*Mb*/$)

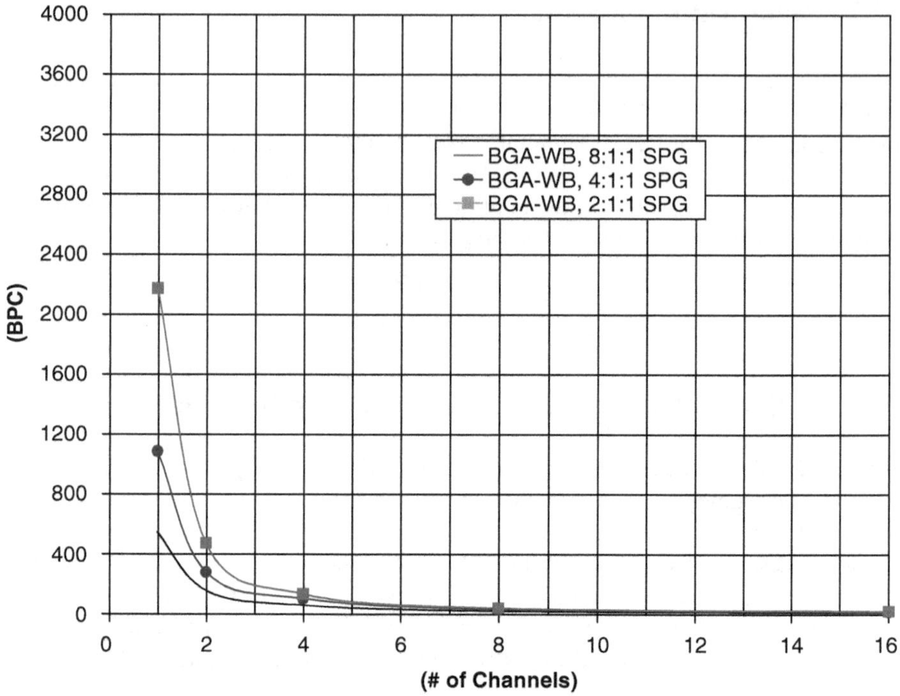

Fig. 12.2 Bandwidth-per-Cost for a BGA-WB package (*Mb/$*)

time, these busses use fewer pins which reduces the overall cost of the design. The value of the BPC metric is that it can be used to select the most cost-effective bus configuration for a particular application.

12.2.2 Results for Ball Grid Array with Wire Bonding

Figure 12.2 shows the *Bandwidth-per-Cost* for the BGA-WB package as signals are added to the segment. This plot again indicates that narrower busses are more cost-effective. This plot also illustrates that the BPC of the BGA-WB is actually less than the QFP-WB package. This is due to the fact that the increase in performance that results from using a BGA (instead of a lead frame) in the level 2 interconnect does not outweigh the increase in cost of the advanced package. This gives rise to the significant conclusion that for certain busses it may be more cost-effective to use a QFP-WB package rather than the more advanced BGA-WB package.

12.2.3 Results for Ball Grid Array with Flip-Chip Bumping

Figure 12.3 shows the *Bandwidth-per-Cost* for the BGA-FC package as a function of the number of signals in the bus segment. This plot again highlights that narrower busses are more cost-effective. Also, the BPC for the BGA-FC is much greater than

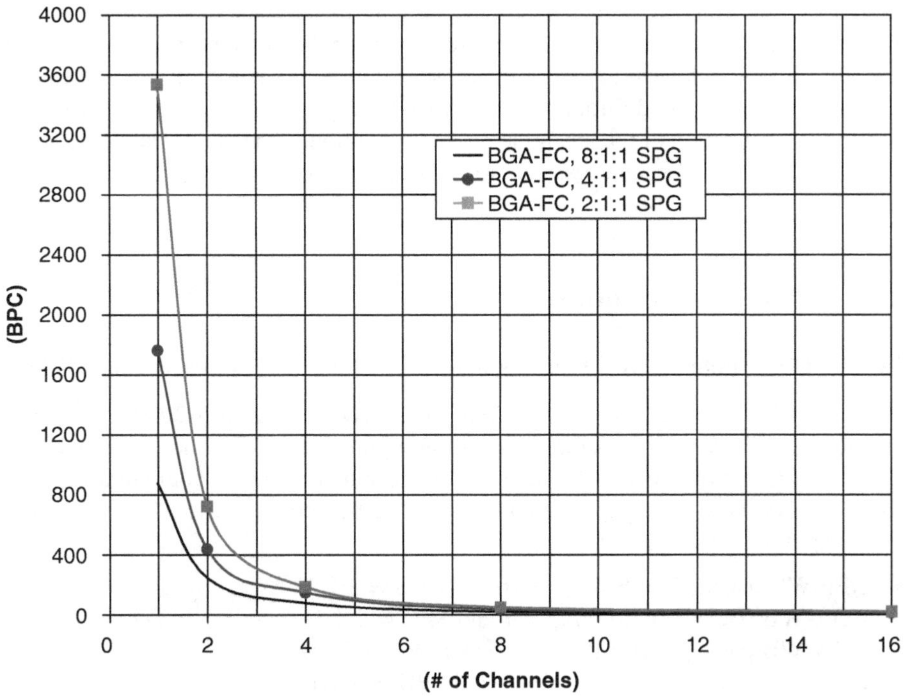

Fig. 12.3 Bandwidth-per-Cost for a BGA-FC package (*Mb*/$)

for either the QFP-WB or the BGA-WB packages. This is due to the fact that the dramatic increase in performance gained by utilizing flip-chip technology outweighs the corresponding increase in cost.

For all the packages studied in this work, it was found that faster, narrower busses were more cost-effective. This trend arises from the negative effects of simultaneously switching signals as channels are added to a bus. These effects make it impractical to simply add I/O pins until the desired throughput is reached. Faster and narrower busses offer the advantage of reduced Simultaneous Switching Noise (SSN), which translates into increased per-pin performance. In addition, since fewer package pins are used, the overall cost of such configurations is lower than that of the wider configuration (which has lower per-pin performance). This trend is being exhibited in industry as more and more computer busses move toward faster, narrower busses. This trend is often referred to as moving from *parallel* to *serial* busses. The unique contribution of this chapter is that it provides a quantitative methodology to find the most cost-effective bus configuration for a given application.

Suppose it is desired to find the most cost-effective package and configuration for a given bus. The performance model of Chap. 11, as well as the BPC model of this chapter can be used to determine if a particular configuration meets the desired performance requirements. The DR and TP parameters can be used to test if a particular configuration meets the desired performance requirements. When two or more configurations meet the performance requirements, their BPC values can next be used to select the most cost-effective configuration. Since these models avoid the use of SPICE, the determination of the most cost-effective bus configuration can be done very quickly. In addition, the BPC metric can provide a quick comparison of the cost-effectiveness of the available packaging options. The analytical models and sizing techniques presented can aid VLSI designers to quickly select the optimal off-chip bus configuration.

12.3 Bus Sizing Example

In order to illustrate the use of the bus sizing technique, consider the example below.

Example 12.1 Consider an on-chip circuit that needs to transmit 2000 $\frac{Mb}{s}$ through the package. Table 12.4 lists the modeled throughput results for the three packages studied in this work. From this table, it is clear that multiple configurations can meet the 2000 $\frac{Mb}{s}$ throughput requirement. The bus configurations that meet the throughput requirement are:

- QFP-WB package with $W_{bus} = 16$ and SPG = 2:1:1
- BGA-WB package with $W_{bus} = 1$ and SPG = 2:1:1
- BGA-WB package with $W_{bus} = 16$ and SPG = 2:1:1
- BGA-FC package with $W_{bus} = 16$ and SPG = 8:1:1
- BGA-FC package with $W_{bus} = 1$ through 16 and SPG = 4:1:1
- BGA-FC package with $W_{bus} = 1$ through 16 and SPG = 2:1:1

Using Table 12.3, it is quickly found that the BGA-WB with $W_{bus} = 1$ and SPG = 2:1:1 is the least expensive configuration that will meet the throughput requirement. This is also apparent in Figs. 12.1, 12.2, and 12.3 which shows that this configuration has a higher BPC than the QFP-WB and BGA-WB configurations. The BGA-FC configurations have higher BPC than the lowest cost BGA-WB configuration. Using the BGA-FC solution has the added advantage that the resulting solution provides additional throughput margin, which is useful in the event that the bus throughput requirement would increase in future revisions of the design.

Table 12.4 Modeled throughput results for packages studied ($\frac{Mb}{s}$)

	Number of channels				
Bus configuration	1	2	4	8	16
QFP-WB 8:1:1	458	324	372	410	820
QFP-WB 4:1:1	916	569	677	777	1554
QFP-WB 2:1:1	1832	959	1182	1430	2860
BGA-WB 8:1:1	553	420	460	498	996
BGA-WB 4:1:1	1106	750	835	935	1871
BGA-WB 2:1:1	2213	1281	1437	1689	3378
BGA-FC 8:1:1	1675	1303	1386	1513	3025
BGA-FC 4:1:1	3349	2272	2446	2814	5628
BGA-FC 2:1:1	6699	3696	4011	4976	9952

Chapter 13
Bus Expansion Encoder

When determining the performance of an off-chip bus, the worst-case noise magnitude must be considered in order to ensure a robust digital system. Chapter 11 presented an analytical model to predict the performance of an off-chip bus using the assumption that each of the noise sources within the package contributes its worst-case noise. Each source of noise within the package (Sect. 8.2) had a particular set of data sequences that resulted in the worst-case noise. This set of sequences dictates the highest performance that the package can achieve. The sequences that result in noise (from *any* noise source) above a certain magnitude can be avoided by designing an encoder which eliminates such sequences. This increases the performance of the package. By inserting this encoder in the signal path on the IC, it ensures that the off-chip data is encoded before traversing the package interconnect. In this way, data sequences which result in noise above a specified limit can be avoided and the bus performance can be increased.

The encoder is constructed to remove any bus sequence which results in a noise event (regardless of the noise source) above a user-specified value. Experimental results demonstrate that this methodology improves the overall performance of the bus even after considering the overhead of the encoder circuit.

The technique maps each element in the original set of on-chip data sequences to an element in an alternate set of sequences (whose noise is bounded by the user-specified limit). Since the alternate set of sequences has a width greater than the width of the original bus, we refer to the resulting circuit as a *bus expansion* encoder. The construction of the encoder/decoder utilizes an implicit, Reduced Ordered Binary Decision Diagram (ROBDD). If it is desired that the noise due to one or more noise sources (described in Sect. 8.2) should be reduced, the method can by employed.

13.1 Constraint Equations

The first step in creating the bus expansion encoder is to create a set of constraint equations. The constraint equations are written so that arbitrary transitions can be evaluated for noise limit violations. When a transition is evaluated using the constraint equations and violates one of the user-defined noise limits, the transition is flagged as *illegal* and is removed from the set of data sequences that are allowed to be

C. Duan et al., *On and Off-Chip Crosstalk Avoidance in VLSI Design*,
DOI 10.1007/978-1-4419-0947-3_13, © Springer Science+Business Media LLC 2010

driven through the package interconnect. Each of the possible off-chip transitions are evaluated against each of the constraint equations. The enumeration of the off-chip transitions is done implicitly, so as to increase the applicability of the technique. After the evaluation is complete, a subset of *legal* transitions remain which are used in the construction of the encoder.

13.1.1 Supply Bounce Constraints

When a pin i in segment j is a V_{DD} pin, it is required that the bounce magnitude due to the electrical parasitics in the package must not exceed the user-defined noise limit P_{supply}. When the pin under evaluation is a V_{DD} pin, a constraint equation is written to determine if any transitions that occur on the bus segment will result in a violation of P_{supply}. The constraint equation takes into account the voltage noise due to the self-inductance of the V_{DD} pin in addition to any mutual inductive or capacitive coupling that occurs due to switching signals in adjacent pins. By multiplying the coupling magnitude by the transition value v_i^j (which can be 0, 1, or -1), the magnitude and sign of the induced noise value is accounted for. This handles the situation for a static pin a static signal pin ($v_i^j = 0$) which has no effect on the noise on the supply pin. The following constraint equation is written for any pin i within a bus segment j that is used as a V_{DD} pin, and is being evaluated for a supply bounce violation:

- $$v_i^j = V_{DD} \Rightarrow P_{supply} \cdot V_{DD} \geq \left(\frac{di}{dt}\right) \cdot \Big[(L_{11}) \cdot (N_1) + \sum_{k=-p_L}^{k=p_L} \big[(M_{1(|k|+1)}) \times \left(v_{i+k}^j\right)\big] + \sum_{k=-p_C}^{k=p_C} \Big[\left(\frac{C_{1(|k|+1)} \cdot Z_0^2}{(0.8)}\right) \cdot \left(v_{i+k}^j\right)\Big]\Big]$$

When a pin i in segment j is a V_{SS} pin it is required that the bounce magnitude due to the electrical parasitics in the package must not exceed the user-defined noise limit P_{gnd}. When the pin under evaluation is a V_{SS} pin, a constraint equation is written to determine if any transitions that occur on the bus segment will result in a violation of P_{gnd}. The following constraint equation is written for any pin within a bus segment j that is used for V_{SS} and is being evaluated for a ground bounce violation:

- $$v_i^j = V_{SS} \Rightarrow P_{gnd} \cdot V_{DD} \geq \left(\frac{di}{dt}\right) \cdot \Big[(L_{11}) \cdot (N_{-1}) + \sum_{k=-p_L}^{k=p_L} \big[(M_{1(|k|+1)}) \times \left(v_{i+k}^j\right)\big] + \sum_{k=-p_C}^{k=p_C} \Big[\left(\frac{C_{1(|k|+1)} \cdot Z_0^2}{(0.8)}\right) \cdot \left(v_{i+k}^j\right)\Big]\Big]$$

13.1.2 Signal Coupling Constraints

13.1.2.1 Glitch Magnitude Constraints

When a pin i in segment j is a signal pin, it is required that the coupled voltage onto that pin does not exceed any of the user-defined noise limits for signal coupling. If the signal pin is static ($v_i^j = 0$), then the glitch magnitude onto the victim pin must not exceed P_0. As in the constraint equations for supply bounce, the magnitude of the coupling contribution of any neighboring pin is multiplied by the transition value v

(which can be 0, 1, or −1) of the neighboring pin. This account for the magnitude and sign of the coupling due to the neighboring pin. The following constraint equation is written for any signal pin within a bus segment j that is static ($v_i^j = 0$) and being evaluated for a glitch violation:

- $v_i^j = 0 \Rightarrow -(P_0 \cdot V_{DD}) \leq \left(\frac{di}{dt}\right) \cdot \left[\sum_{k=-p_L}^{k=p_L} \left[(M_{1(|k|+1)}) \cdot (v_{i+k}^j) \right] + \sum_{k=-p_C}^{k=p_C} \left[\left(\frac{C_{1(|k|+1)} \cdot Z_0^2}{(0.8)} \right) \cdot (v_{i+k}^j) \right] \right] \leq (P_0 \cdot V_{DD})$

13.1.2.2 Risetime Degradation Constraints

When a signal pin i in segment j transitions from a logic 0 to a logic 1 ($v_i^j = 1$), it is required that the coupled voltage onto that pin does not hinder its risetime. In this situation the cumulative value of the mutual coupling can be exploited to actually aid the transition on the victim signal pin. By requiring that the cumulative coupled voltage on the victim pin either equals or exceeds the user-defined noise limit P_1, it is also possible to improve the risetime of the victim signal (i.e., it is possible to speed up the victim signal). By setting P_1 to zero, it is guaranteed that the risetime for the victim signal is not hindered. The following constraint equation is written for any signal pin within a bus segment j that is undergoing a rising transition ($v_i^j = 1$) and is being evaluated for a rising edge degradation violation:

- $v_i^j = 1 \Rightarrow P_1 \cdot V_{DD} \leq \left(\frac{di}{dt}\right) \cdot \left[\sum_{k=-p_L}^{k=p_L} \left[(M_{1(|k|+1)}) \cdot (v_{i+k}^j) \right] + \sum_{k=-p_C}^{k=p_C} \left[\left(\frac{C_{1(|k|+1)} \cdot Z_0^2}{(0.8)} \right) \cdot (v_{i+k}^j) \right] \right]$

13.1.2.3 Falltime Degradation Constraints

In a similar manner, when a signal pin i in segment j transitions from a logic 1 to a logic 0 ($v_i^j = -1$), it is required that the coupled voltage onto that pin does not hinder its falltime. By requiring that the cumulative coupled voltage on the victim pin either equals or exceeds the user-defined noise limit P_{-1}, it is also possible to improve the falltime of the victim signal (i.e., it is possible to speed up the victim signal). By setting P_{-1} to zero, it is guaranteed that the falltime for the victim signal is not hindered. The following constraint equation is written for any signal pin within a bus segment j that is undergoing a falling transition ($v_i^j = -1$) and being evaluated for a falling edge degradation violation:

- $v_i^j = -1 \Rightarrow P_{-1} \cdot V_{DD} \geq \left(\frac{di}{dt}\right) \cdot \left[\sum_{k=-p_L}^{k=p_L} \left[(M_{1(|k|+1)}) \cdot (v_{i+k}^j) \right] + \sum_{k=-p_C}^{k=p_C} \left[\left(\frac{C_{1(|k|+1)} \cdot Z_0^2}{(0.8)} \right) \cdot (v_{i+k}^j) \right] \right]$

13.1.3 Capacitive Bandwidth Limiting Constraints

When a pin i in segment j is a signal pin, it is required that the capacitive bandwidth limitation due to adjacent switching signal pins does not degrade the *original* risetime by more than P_{BW}. As described in Sect. 8.2.3, the capacitance that a victim signal will experience will depend on the self-capacitance (C_0) in addition to mutual coupling capacitance to neighboring signal pins (C_{1k}). The original risetime is defined as the risetime achieved when all aggressing signals within the segment are static. In this case, the mutual capacitance to all other signal pins will be $1 \cdot C_{1k}$. When the neighboring aggressor pins switch, the resulting risetime of the victim can be either improved, degraded, or unhindered (depending on the switching pattern of the aggressor pins).

The worst-case transition on the bus (which causes the largest risetime degradation on the victim) will be when all aggressor signal pins switch in the opposite direction as the victim signal. This results in a doubling of the mutual capacitance to all neighboring pins due to the doubling of the voltage excursion between signals. This makes the capacitance to aggressing signal pins $2 \cdot C_{1k}$. The best-case transition on the bus (which causes the largest risetime improvement on the victim) will be when all signal pins switch in the same direction. In this case the coupling capacitance between pins will be zero ($0 \cdot C_{1k}$) because there is no net voltage difference across the pins. Since P_{BW} is described as a percentage of risetime degradation, the constraint equation must account for the *original* risetime in addition to the risetime which results when neighboring signals switch. The following constraint equation is written for any signal pin within a bus segment j that is undergoing a positive transition ($v_i^j = 1$) and being evaluated for a capacitive bandwidth limitation violation:

- $v_i^j = 1 \Rightarrow$

$$P_{BW} \leq \frac{\left[2.2 \cdot \left(C_0 + \sum_{k=-pc}^{k=pc} C_{1(|k|+1)}\right) \cdot Z_0\right] - \left[2.2 \cdot \left(C_0 + \sum_{k=-pc}^{k=pc} C_{1(|k|+1)} \cdot \left(1 - v_{i+k}^j\right)\right) \cdot Z_0\right]}{\left[2.2 \cdot \left(C_0 + \sum_{k=-pc}^{k=pc} C_{1(|k|+1)}\right) \cdot Z_0\right]}$$

In the above constraint, multiplying the aggressor coupling capacitance by $(1 - v_{i+k}^j)$ handles the contribution of the mutual capacitance to the bandwidth limitation. Since the aggressor signal pin can take on values of $v_{i+k}^j \in \{-1, 0, 1\}$, this evaluates to $(1 - v_{i+k}^j) \in \{2, 1, 0\}$. This quantity is multiplied with C_{1k} and then added to C_0 to give the total amount of capacitance that the victim pin charges.

In a similar manner, the bandwidth limitation constraint can be written for a victim signal that undergoes a falling transition ($v_i^j = -1$). In this case, the contribution of the coupling capacitance is handled by multiplying the mutual capacitance by $(| -1 - v_{i+k}^j|)$. Using this expression, the aggressor signal transition values ($v_{i+k}^j \in \{-1, 0, 1\}$) evaluate to $(| -1 - v_{i+k}^j|) \in \{0, 1, 2\}$. The following constraint equation is written for any signal pin i within a bus segment j that undergoes a falling transition ($v_i^j = -1$) and is being evaluated for a capacitive bandwidth limitation violation:

- $v_i^j = -1 \Rightarrow$

$$P_{BW} \geq \frac{\left[2.2 \cdot \left(C_0 + \sum_{k=-pc}^{k=pc} C_{1(|k|+1)} \cdot (|-1 - v_{i+k}^j|)\right) \cdot Z_0\right] - \left[2.2 \cdot \left(C_0 + \sum_{k=-pc}^{k=pc} C_{1(|k|+1)}\right) \cdot Z_0\right]}{\left[2.2 \cdot \left(C_0 + \sum_{k=-pc}^{k=pc} C_{1(|k|+1)}\right) \cdot Z_0\right]}$$

13.1.4 Impedance Discontinuity Constraints

When a pin i in segment j is a signal pin it is required that the magnitude of the reflected voltage due to impedance discontinuities within the package must not exceed P_Γ. As described in Sect. 8.2.5, the impedance of a given signal path will depend on its self-inductance (L_{11}) and self-capacitance (C_0) in addition to any mutual inductance (M_{1k}) and mutual capacitance (C_{1k}). This means that the data sequences present on neighboring signal pins will alter the effective inductance and capacitance of the victim signal. This directly effects the characteristic impedance of the victim pin, which is used in the evaluation of reflected noise. This must be accounted for in the constraint equation. By multiplying this additional mutual inductance and capacitance values by the transition values of the aggressor neighbors (v_{i+k}^j), the polarity and cumulative effect of the coupling can be handled. The following constraint equation is written for any signal pin i within a bus segment j that undergoes a positive transition ($v_i^j = 1$) and is being evaluated for an impedance discontinuity violation:

- $$v_i^j = 1 \Rightarrow P_\Gamma \leq \left| \frac{\sqrt{\dfrac{L_{11} + \sum_{k=-pL}^{k=pL} (M_{1(|k|+1)}) \cdot (v_{i+k}^j)}{C_0 + \sum_{k=-pC}^{k=pC} (C_{1(|k|+1)}) \cdot (v_{i+k}^j)}} - Z_0}{\sqrt{\dfrac{L_{11} + \sum_{k=-pL}^{k=pL} (M_{1(|k|+1)}) \cdot (v_{i+k}^j)}{C_0 + \sum_{k=-pC}^{k=pC} (C_{1(|k|+1)}) \cdot (v_{i+k}^j)}} + Z_0} \right|$$

In a similar manner, when a pin i in segment j is a signal pin and undergoes a falling transition ($v_i^j = -1$), it is required that the magnitude of the reflected voltage due to impedance discontinuities within the package must not exceed P_Γ. In this case the contribution of the mutually coupled inductance and capacitance must be inverted to properly represent its cumulative nature. This is accomplished by multiplying the mutual inductance and capacitance values by the negative of the transition value ($-v_{i+k}^j$), thereby accounting for the polarity of the mutually coupled voltage. Note that the polarity of Γ is reversed when the forcing function is a falling edge [97]. The following constraint equation is written for any signal pin i within a bus segment j that undergoes a falling transition ($v_i^j = -1$) and is being evaluated for an impedance discontinuity violation:

- $v_i^j = -1 \Rightarrow P_\Gamma \leq \left| \frac{\sqrt{\frac{L_{11} + \sum_{k=-p_L}^{k=p_L} (M_{1(|k|+1)}) \cdot \left(-v_{i+k}^j\right)}{C_0 + \sum_{k=-p_C}^{k=p_C} (C_{1(|k|+1)}) \cdot \left(-v_{i+k}^j\right)}} - Z_0}{\sqrt{\frac{L_{11} + \sum_{k=-p_L}^{k=p_L} (M_{1(|k|+1)}) \cdot \left(-v_{i+k}^j\right)}{C_0 + \sum_{k=-p_C}^{k=p_C} (C_{1(|k|+1)}) \cdot \left(-v_{i+k}^j\right)}} + Z_0} \right|$

13.1.5 Number of Constraint Equations

For each V_{DD} pin within a segment, one equation is written that represents the supply bounce constraint (P_{supply}). For each V_{SS} pin within a segment, one equation is written that represents the ground bounce constraint (P_{gnd}). For each signal pin, constraint equations are written for the three transition values that the pin can have ($v_i^j \in \{-1, 0, 1\}$). When a signal pin is static ($v_i^j = 0$), one equation must be written to constrain the glitching noise due to switching neighbor pins (P_0). When a signal pin transitions high ($v_i^j = 1$), three equations must be written that account for rising edge degradation (P_1), capacitive bandwidth limitation (P_{BW}), and impedance discontinuities (P_Γ). When a signal pin transitions low ($v_i^j = -1$), three more equations must be written that account for falling edge degradation (P_{-1}), capacitive bandwidth limitation (P_{BW}), and impedance discontinuities (P_Γ). This gives the total number of constraint equations that are written as:

$$N_{constraints} = N_p + N_g + (7 \cdot W_{bus}) \tag{13.1}$$

The number of constraints can be reduced for particular applications. For a particular package, some noise sources may be negligible and can be ignored in the constraint evaluation. For the three packages studied in this work, it was found that in all cases the supply bounce and inductive signal coupling were the dominant noise sources. This is a consequence of noise being dominated by the inductance within the package interconnect. If only the inductive noise components are considered, then each signal pin needs only three constraint equations: P_0, P_1, and P_{-1}. This reduces the number of constraint equations to:

$$N_{ind\text{–}constraints} = N_p + N_g + (3 \cdot W_{bus}) \tag{13.2}$$

13.1.6 Number of Constraint Evaluations

Each of the constraint equations must be evaluated against each of the possible transitions that are to be transmitted through the package. Since each signal pin can take on three unique transition values ($v_i^j \in \{-1, 0, 1\}$), then the total number of possible transitions for a given bus segment is bounded by:

$$N_{transitions} = (W_{bus} + 2 \cdot \max(p_L, p_C))^3 \tag{13.3}$$

In practice, however, some of the signals in the neighborhood of the signal of interest may be supply pins, resulting in fewer constraints. This gives the total number of constraint equation evaluations as:

$$N_{evaluations} = (N_{constraints}) \cdot (N_{transitions}) \tag{13.4}$$

Again, the number of evaluations can be reduced by limiting the noise sources that are considered. It should be noted that due to the symmetry of the bus segment representation (which mimics the regular manner in which the busses are implemented in practice), the evaluations need only be performed on the signals within a single bus segment and not on the entire bus. This symmetry allows a dramatic reduction in computation time compared to the case where the entire bus is analyzed monolithically.

13.2 Encoder Construction

Once the transitions have been tested against all of the constraining equations, a subset of legal transitions remain from the original set of possible transitions. Assume that the original set of constraints were written for a bus segment with n signals. Using the legal transitions remaining, we next find the maximum effective bus size m, such that the transitions on this bus can be encoded using the remaining legal transitions on the physical n-bit bus.

13.2.1 Encoder Algorithm

From the set of legal vector sequences, we next create a ROBDD [20, 22] G, to encode legal bus transitions. We then find the effective size m of the bus that can be encoded using the transitions in G, using a ROBDD based algorithm. Note that the ROBDD G has $2n$ variables. The first n variables refer to the *from* vertices and the next n variables refer to the *to* vertices of the vector transition. There is a legal edge between vertices v_1 and v_2 iff $G(v_1, v_2) = 1$.

Note that for a vector sequence v^j, we can construct minterms in G to encode transitions between vectors w^j_{from} and w^j_{to}. The end-points of this edge (w^j_{from} and w^j_{to}) can be constructed given v^j, as follows:

- $w^j_{from,i} = 0$ if $v^j_i = 1$ (i.e. the signal is rising) or if $v^j_i = 0$ (i.e. the signal is static).
- $w^j_{from,i} = 1$ if $v^j_i = -1$ (i.e. the signal is falling) or if $v^j_i = 0$ (i.e. the signal is static).

Similarly, we can write

- $w^j_{to,i} = 1$ if $v^j_i = 1$ or if $v^j_i = 0$.
- $w^j_{to,i} = 0$ if $v^j_i = -1$ or if $v^j_i = 0$.

$G(w^j_{from}, w^j_{to}) = 1$ indicates the legality (from an inductive cross-talk viewpoint) of the transition from vector w^j_{from} to w^j_{to}. Therefore, given a set of vector sequences $\{v^j\}$ which are *legal*, we can construct a ROBDD G whose minterms $(w_{from} : w_{to})$ are vectors in B^{2n}, such that they indicate a legal transition (from an inductive cross-talk viewpoint) between the source (w_{from}) and sink (w_{to}) vertices. Note that the ":" symbol above refers to the concatenation operator.

If an m-bit bus can be encoded using the legal transitions in G, then there must exist a set of vertices $V_c \subseteq B^n$ such that:

- Each $v_s \in V_c$ has at least 2^m outgoing edges $e(v_s, v_d)$ (including the self edge), such that the destination vertex $v_d \in V_c$.
- The cardinality of V_c is at least 2^m.

The resulting encoder is memory based. Note that the physical size of the bus n is obviously greater than or equal to m.

Given G, we find m using Algorithm 13.1. The input to the algorithm is m and G. We first find the out-degrees (self-edges are counted) of each $v_s \in B^n$. This is done by logically ANDing the ROBDD of the vertex v_s with G. We find the cardinality of the resulting ROBDD – it represents the out-degree of v_s. If the number of out-edges of any v_s is greater than 2^m, we add v_s (and its out-degree) into a hash table V.

For each $v_s \in V$, we next check if each of its destination nodes v_d are in V. If $v_d \notin V$, we decrement the out-degree of v_s by 1. If the out-degree of v_s becomes less than 2^m, we remove v_s from V. These operations are performed until convergence. If at this point, the number of surviving vertices in V is 2^m or more, then an m-bit memoryless CODEC can be constructed from G.

We initially call the algorithm with $m = n - 1$ (where n is the physical bus size). If an m bit bus cannot be encoded using G, then we decrement m. We repeat this until we find a value of m such that the m-bit bus can be encoded by G.

Algorithm 13.1 Testing if G can encode an m-bit bus

```
test_encoder(m, G)
find out − degree(v_s) of each node v_s, insert (v_s, out − degree(v_s)) in V if out −
degree(v_s) ≥ 2^m
degrees_changed = 1
while degrees_changed do
   degrees_changed = 0
   for each v_s ∈ V do
      for each v_d S.T. G(v_s, v_d) = 1 do
         if v_d ∉ V then
            decrement out − degree(v_s) in V
            degrees_changed = 1
         end if
         if out − degree(v_s) < 2^m then
            V ← V \ v_s
            break
         end if
```

```
      end for
    end for
  end while
  if |V| ≥ 2^m then
    print(m-bit bus may be encoded using G)
  else
    print(m-bit bus cannot be encoded using G)
  end if
```

For a memory-less encoder, the subgraph induced by the edges in V_c should be a clique. Memory-based encoders are explored in this work, due to the fact that they allow effective widths which are higher than memory-less encoders.

Note that this entire analysis needs to be performed for a representative bus segment. In other words, even if the bus is very wide, the analysis is performed for a single segment, which is typically very small. This segment could be part of a much larger bus, and the analysis would be valid for all segments of the bus.

13.2.2 Encoder Overhead

The end result of the encoder construction is a mapping between transitions on an m-bit on-chip bus to transitions on an n-bit off-chip bus. The transitions on the n-bit off-chip bus are selected so that the worst-case noise resulting from any of these transitions is less than or equal to the user-specified noise limit. Since the transitions on the n-bit bus avoid any noise greater than the user-specified limit, they can be transmitted at a higher datarate The net improvement must also account for the overhead in the number of bits utilized for the bus (which is $n-m$). The computation of the overhead for the bus expansion encoding technique is shown in Eq. 13.5.

$$Overhead_{bus\text{-}expansion} = \left(\frac{n-m}{n}\right) \cdot 100 \tag{13.5}$$

13.3 Decoder Construction

The decoder construction is identical to the encoder construction however the process is performed in reverse. The combinational logic for the decoder is simply a reverse permutation of the logic for the encoder. In the case of the memory-less decoder, the logic is induced by that of the encoder. This is also the case for the memory-based decoders.

13.4 Experimental Results

This section presents the experimental results that were performed to verify the effectiveness of the bus expansion encoder.

13.4.1 3-Bit Fixed $\frac{di}{dt}$ Example

The first example is a 3-bit bus that is evaluated for a fixed $\frac{di}{dt}$. This example illustrates how the encoding technique directly reduces noise within the package. A $\frac{di}{dt}$ of 33 $\frac{MA}{s}$ was chosen that corresponds to a datarate of 550 $\frac{Mb}{s}$ in a 50 Ω system using Eq. 11.6. The electrical parameters for the BGA-WB package are used from Table 9.1. The bus configuration has three signal pins ($W_{bus} = 3$), one V_{DD} pin ($N_p = 1$), and one V_{SS} pin ($N_g = 1$) for a total of 5 pins within the bus segment ($N_{segment} = 5$). This configuration yields a *Signal-to-Power-Ground* metric of $SPG = 3{:}\ 1{:}\ 1$ and a *Signal-to-Power-Ground Ratio* of $SPGR = 3$. The mutual inductive and capacitive coupling is considered to the 2 most adjacent pins ($p_L = p_C = 2$). This simplifies the analysis by ignoring inductive coupling less than 15% and capacitive coupling less than 50 fF. The bus configuration is shown in Fig. 13.1.

This bus was encoded using two sets of constraints: *aggressive* (P_{supply}, P_{gnd}, P_0, P_1, P_{-1}, P_{BW}, and P_Γ set to 5%) and *non-aggressive* (P_{supply}, P_{gnd}, P_0, P_1, P_{-1}, P_{BW}, and P_Γ set to 10%). The first step in the bus expansion technique is to write the constraint equations. Equation 13.1 indicates that 24 constraint equations need to be written for this package; however, from the results in Chap. 11, it is known that the BGA-WB performance is dominated by the inductive nature of its interconnect. This allows the number of constraint equations to be reduced to 11 using Eq. 13.2. The following lists the 11 constraint equations for this example. These constraints have been simplified by removing terms with $v_i^j = 0$.

1) $v_0^j = V_{DD} \Rightarrow P_{supply} \cdot V_{DD} \leq \frac{di}{dt} \cdot \left[(L_{11} \cdot N_1) + \left(M_{13} \cdot v_3^{j-1}\right) + \left(M_{12} \cdot v_1^j\right) + \left(M_{13} \cdot v_2^j\right)\right]$

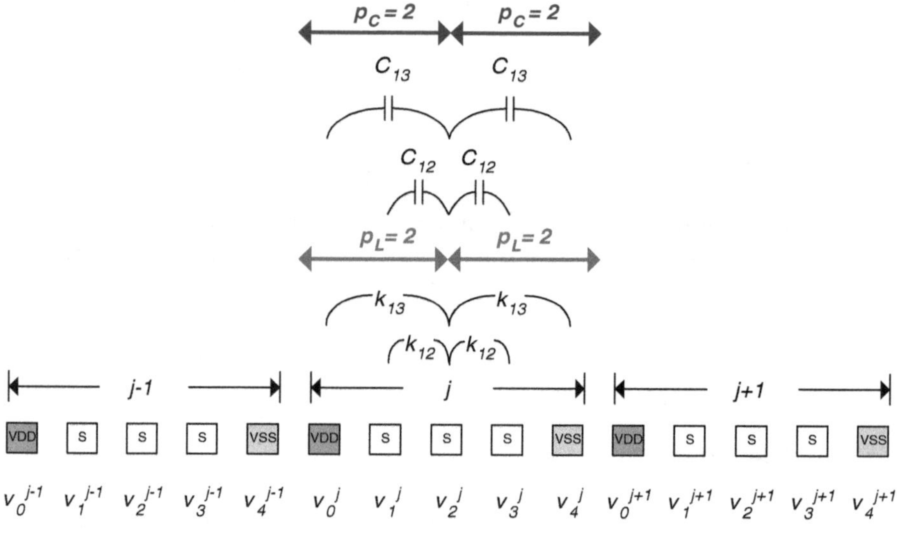

Fig. 13.1 3-Bit bus example

2) $v_1^j = 1 \Rightarrow P_1 \cdot V_{DD} \leq \frac{di}{dt} \cdot \left[\left(M_{12} \cdot v_2^j\right) + \left(M_{13} \cdot v_3^j\right)\right]$

3) $v_1^j = -1 \Rightarrow P_{-1} \cdot V_{DD} \geq \frac{di}{dt} \cdot \left[\left(M_{12} \cdot v_2^j\right) + \left(M_{13} \cdot v_3^j\right)\right]$

4) $v_1^j = 0 \Rightarrow -(P_0 \cdot V_{DD}) \leq \frac{di}{dt} \cdot \left[\left(M_{12} \cdot v_2^j\right) + \left(M_{13} \cdot v_3^j\right)\right] \leq (P_0 \cdot V_{DD})$

5) $v_2^j = 1 \Rightarrow P_1 \cdot V_{DD} \leq \frac{di}{dt} \cdot \left[\left(M_{12} \cdot v_1^j\right) + \left(M_{12} \cdot v_3^j\right)\right]$

6) $v_2^j = -1 \Rightarrow P_{-1} \cdot V_{DD} \geq \frac{di}{dt} \cdot \left[\left(M_{12} \cdot v_1^j\right) + \left(M_{12} \cdot v_3^j\right)\right]$

7) $v_2^j = 0 \Rightarrow -(P_0 \cdot V_{DD}) \leq \frac{di}{dt} \cdot \left[\left(M_{12} \cdot v_1^j\right) + \left(M_{12} \cdot v_3^j\right)\right] \leq (P_0 \cdot V_{DD})$

8) $v_3^j = 1 \Rightarrow P_1 \cdot V_{DD} \leq \frac{di}{dt} \cdot \left[\left(M_{13} \cdot v_1^j\right) + \left(M_{12} \cdot v_2^j\right)\right]$

9) $v_3^j = -1 \Rightarrow P_{-1} \cdot V_{DD} \geq \frac{di}{dt} \cdot \left[\left(M_{13} \cdot v_1^j\right) + \left(M_{12} \cdot v_2^j\right)\right]$

10) $v_3^j = 0 \Rightarrow -(P_0 \cdot V_{DD}) \leq \frac{di}{dt} \cdot \left[\left(M_{13} \cdot v_1^j\right) + \left(M_{12} \cdot v_2^j\right)\right] \leq (P_0 \cdot V_{DD})$

11) $v_4^j = V_{SS} \Rightarrow P_{gnd} \cdot V_{DD} \leq \frac{di}{dt} \cdot \left[(L_{11} \cdot N_{-1}) + \left(M_{13} \cdot v_2^j\right) + \left(M_{12} \cdot v_3^j\right) + \left(M_{13} \cdot v_1^{j+1}\right)\right]$

All possible transitions on the bus are then evaluated using the 11 constraint equations. From this evaluation, transitions that violate any of the constraints are flagged as *illegal* and removed from subset of *legal* transitions used in the encoder construction. Table 13.1 shows the results of the constraint evaluations. In this table, if one of the

Table 13.1 Constraint evaluations for 3-Bit, fixed $\frac{di}{dt}$ bus expansion example

Original transition	Aggressive	Non-aggressive
000	000	000
001	001	001
00-1	00-1	00-1
010	010	010
011	Violates 1,4	011
01-1	01-1	01-1
0-10	0-10	0-10
0-11	0-11	0-11
0-1-1	Violates 4,11	0-1-1
100	100	100
101	Violates 1,7	101
10-1	10-1	10-1
110	Violates 1,10	110
111	Violates 1,2,5,8	Violates 11
11-1	Violates 1	11-1
1-10	1-10	1-10
1-11	Violates 1	1-11
1-1-1	Violates 11	1-1-1
-100	-100	-100
-101	-101	-101
-10-1	Violates 7,11	-10-1
-110	-110	-110
-111	Violates 1	-111
-11-1	Violates 11	-11-1
-1-10	Violates 10,11	-1-10
-1-11	Violates 11	-1-11
-1-1-1	Violates 3,6,9,11	Violates 1

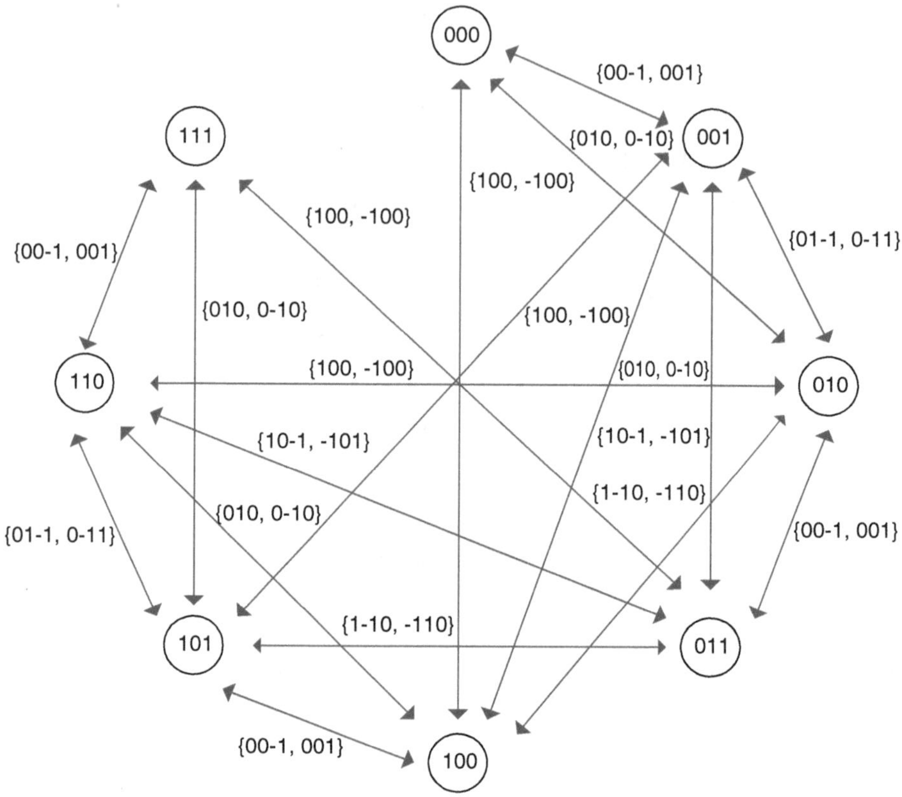

Fig. 13.2 Directed graph for the 3-Bit, fixed $\frac{di}{dt}$ bus expansion example

original transitions violates a constraint equation, the number(s) of the constraint equation is listed.

After all transitions have been evaluated against the constraints, the remaining legal transitions are used to create a directed graph. Figure 13.2 shows the directed graph that was created from the non-aggressive constraint evaluation. This figure does not show the self-edges for simplicity.

From the directed graphs for both the non-aggressive and aggressive constraints, the largest effective bus width m was constructed, such that the noise for this bus is within the user-defined limits. From the mapping between the transitions in the m-bit bus to those in the legal transitions in the n-bit bus, the encoder state machine is constructed. Figure 13.3 shows the overhead for the bus expansion encoder (Eq. 13.5) for both the aggressive and non-aggressive constraints. This figure suggests that the overhead of the encoder approaches an asymptotic limit as the number of channels are added to the bus segment.

SPICE simulations were performed to quantify the noise reductions achieved by the encoder. Figure 13.4 shows the ground bounce reduction achieved by the encoder for both the aggressive and non-aggressive constraints. This figure shows

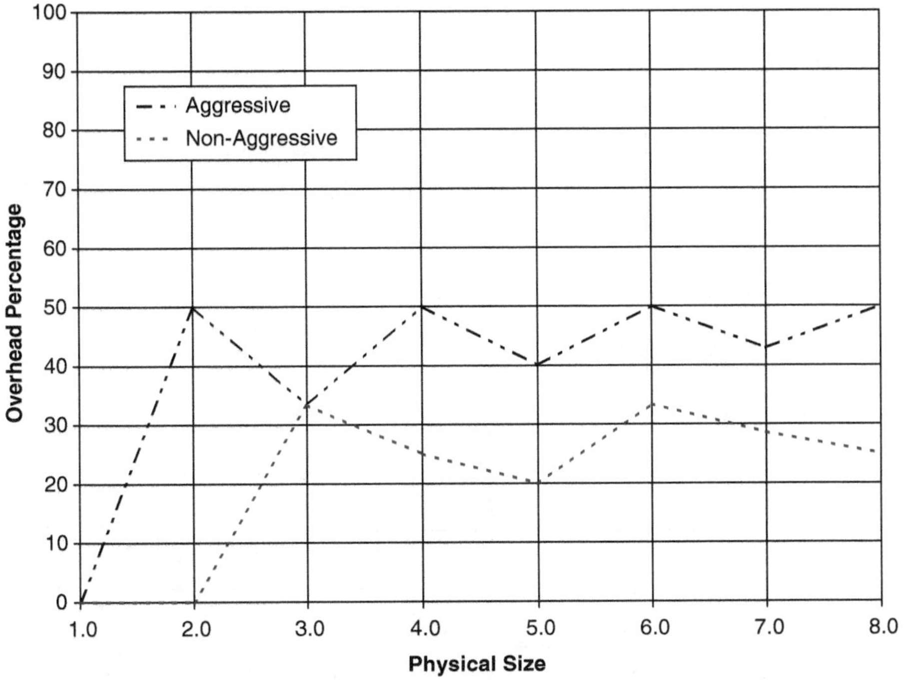

Fig. 13.3 Bus expansion encoder overhead for the fixed $\frac{di}{dt}$ example

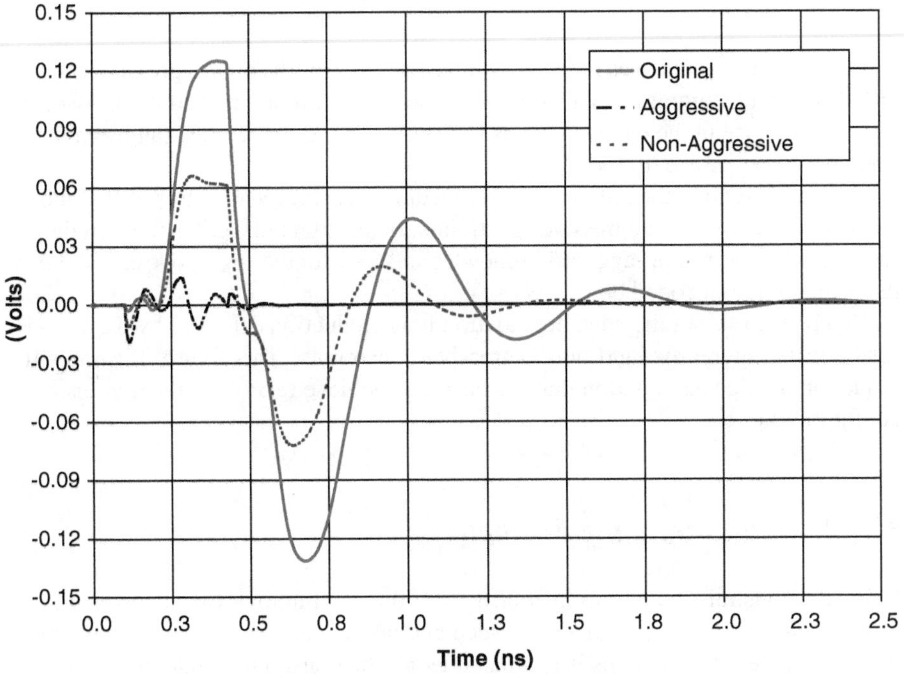

Fig. 13.4 SPICE simulation of ground bounce for 3-Bit, fixed $\frac{di}{dt}$ example

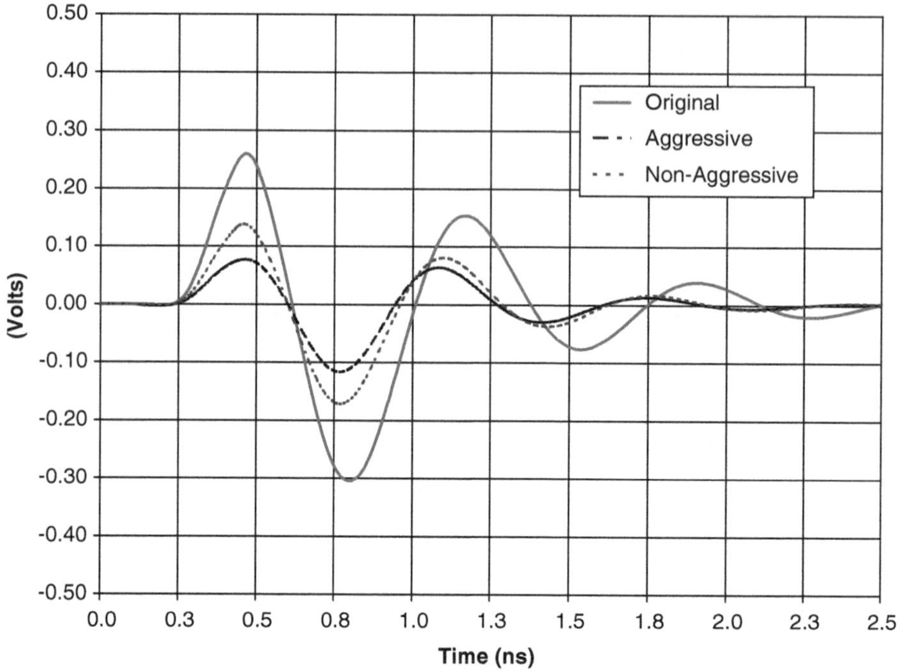

Fig. 13.5 SPICE simulation of glitching noise for 3-Bit, fixed $\frac{di}{dt}$ example

that the ground bounce was reduced as much as 50% for the non-aggressive encoder and 89% for the aggressive encoder (over the original non-encoded bus). Also, the worst-case noise in the encoded bus is just within the user-specified (aggressive and non-aggressive) noise limits.

Figure 13.5 shows the glitch magnitude reduction achieved by the encoder for both the aggressive and non-aggressive constraints. In this figure the glitch magnitude was reduced 44% for the non-aggressive encoder and 68% for the aggressive encoder over the original non-encoded bus.

Figure 13.6 shows the edge degradation noise reduction achieved by the encoder for both the aggressive and non-aggressive constraints. This figure illustrates the reduction in edge degradation that the encoders achieve relative to the non-encoded configuration.

13.4.2 3-Bit Varying $\frac{di}{dt}$ Example

Using the constraint equations in Sect. 13.4.1, the maximum $\frac{di}{dt}$ that the package can achieve can be found for both the encoded and non-encoded cases. For this example the original set of transitions is evaluated using the constraint equations to find the maximum $\frac{di}{dt}$ and, in turn, the maximum per-pin datarate for the non-encoded bus.

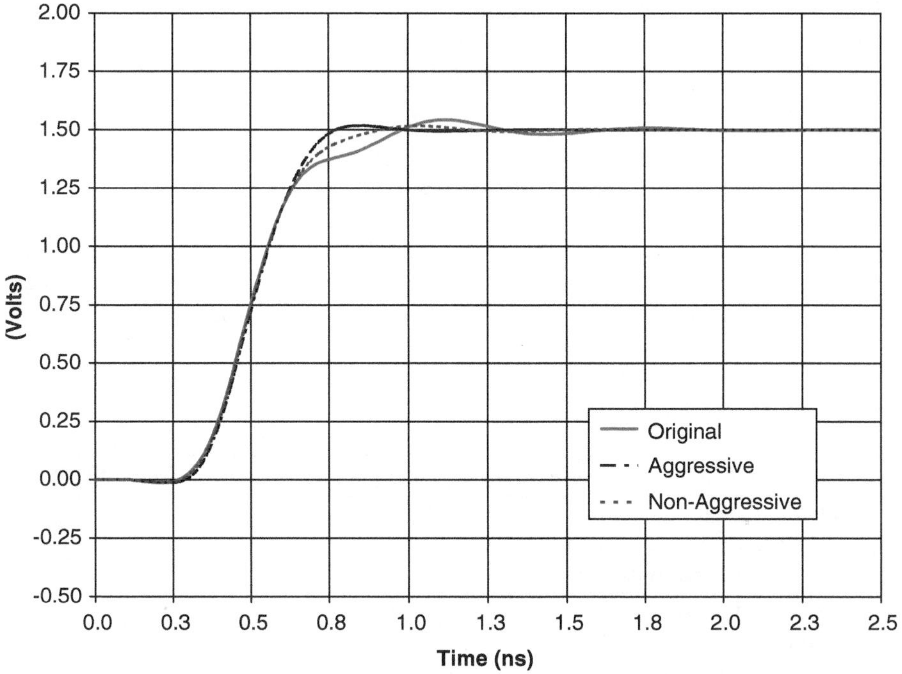

Fig. 13.6 SPICE simulation of edge coupling reduction for 3-Bit, fixed $\frac{di}{dt}$ example

For the non-encoded evaluation, no vectors are removed from the original set; instead, $\frac{di}{dt}$ is increased to its maximum value without violating any of the constraint equations. At that point, this $\frac{di}{dt}$ indicates the fastest datarate that the non-encoded bus can achieve.

For the encoded evaluation, the reduced transition set from Sect. 13.4.1 is evaluated in the same manner. In the encoded case, the worst-case transitions have already been removed from the legal transition set and will result in a faster $\frac{di}{dt}$ that can be achieved without violating any of the constraints. Table 13.2 shows the results for

Table 13.2 Experimental results for the 3-Bit, varying $\frac{di}{dt}$ example

	Original	Encoded
Max di/dt	13.3 MA/s	37 MA/s
Max Datarate	222 Mb/s	617 Mb/s
# of Legal Transitions	27	13
Physical Size of Bus	3	3
Effective Size of Bus	3	2
Encoder Overhead	—	33%
Bus Throughput	666 Mb/s	1234 Mb/s
Throughput Improvement	—	46%

this example. In this table the maximum $\frac{di}{dt}$ for the original and encoded cases are reported. This table illustrates that even after considering the overhead of the encoder, the bus throughput is increased as much as 46% *using the same three physical signal pins* on the package.

13.4.3 Functional Implementation

The bus expansion encoders were implemented to verify their functionality and feasibility. The Verilog hardware description language (HDL) was used to create the encoder circuitry [29, 78]. The Verilog implementation consisted of a series of *CASE* statement that mapped the on-chip m-bit transitions into the n-bit transitions that could be transmitted through the package. Algorithm 13.2 shows the Verilog HDL code for the bus expansion encoder.

Algorithm 13.2 Verilog implementation of m-Space to n-Space encoder mapping

```
module expansion_encoder (expansion_data_out, data_in);
 output reg [n-1:0] expansion_data_out;
 in         [m-1:0] data_in;
  always @ (data_in)
    begin
      case (data_in)
        m'bxx : expansion_data_out = n'bxxx;
          :
          :
        m'bxx : expansion_data_out = n'bxxx;
      endcase
    end
endmodule
```

Figures 13.7–13.10 show the Verilog simulation results for aggressively constrained bus expansion encoders for effective sizes of 2, 4, 6, and 8. For each waveform the m-bit clock and data are plotted in addition to the encoded n-bit data which is transmitted off-chip.

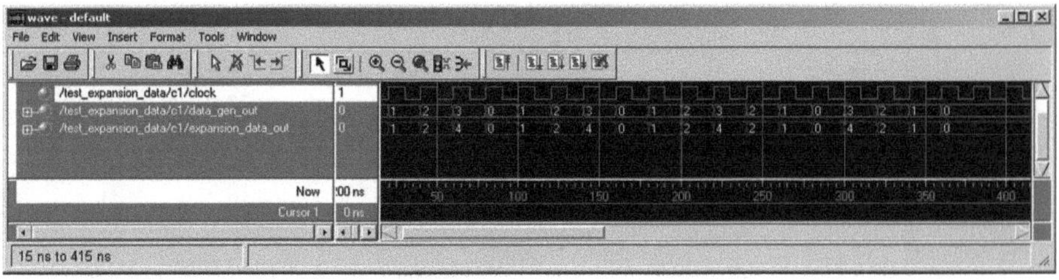

Fig. 13.7 Verilog simulation results for a 2-Bit bus expansion encoder

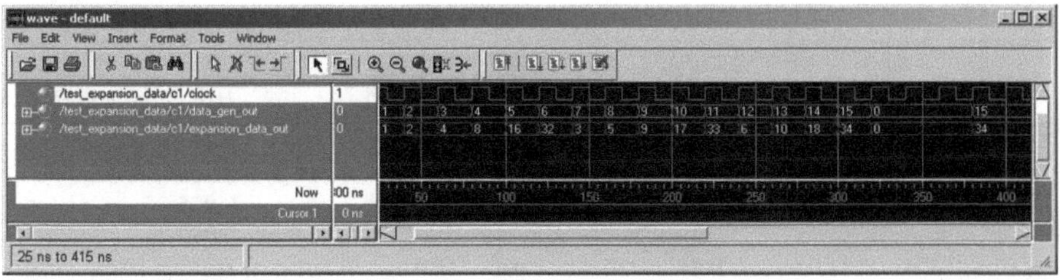

Fig. 13.8 Verilog simulation results for a 4-Bit bus expansion encoder

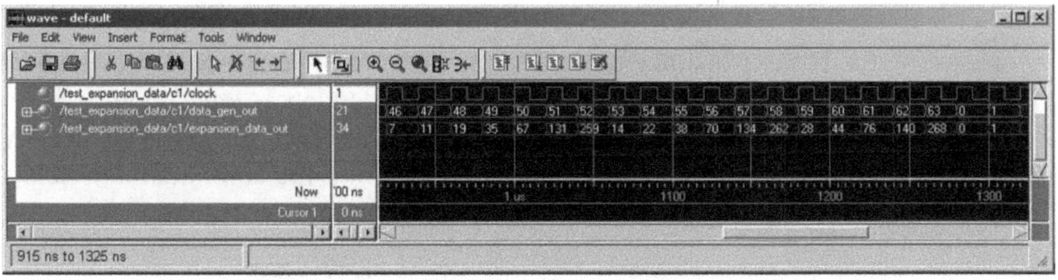

Fig. 13.9 Verilog simulation results for a 6-Bit bus expansion encoder

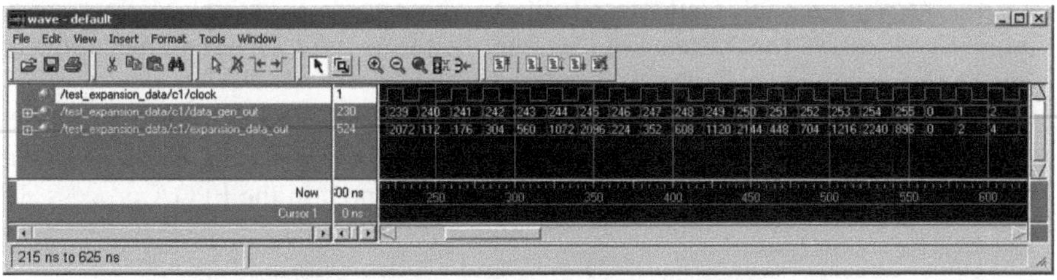

Fig. 13.10 Verilog simulation results for a 8-Bit bus expansion encoder

13.4.4 Physical Implementation

In order to evaluate the feasibility of the expansion encoder/decoder, the physical design of the circuitry was performed.

13.4.4.1 TSMC 0.13 μm ASIC Process

The encoders were synthesized using the TSMC 0.13 μm CMOS IC process to understand the impact on delay and area if the encoders were integrated on-chip. Encoders for effective bus sizes of 2, 4, 6, and 8 were implemented. For each of these sizes, both the aggressive (5% of V_{DD}) and the non-aggressive (10% of V_{DD}) encoders were synthesized. Table 13.3 lists the delay and area impact of the bus expansion encoders when implemented in a TSMC 0.13 μm process. This table illustrates that

Table 13.3 Bus expansion encoder synthesis results in a TSMC 0.13 μm process

	Bus size (m)	Noise limit	
	—	5% (aggressive)	10% (non-aggressive)
Delay (ns)	2	0.170	encoder not required
	4	0.670	0.503
	6	1.150	0.955
	8	1.310	0.983
Area (um^2)	2	22	encoder not required
	4	152	114
	6	614	509
	8	1,181	886

incorporating the bus expansion encoder in a modern VLSI design results in a negligible area and delay penalty. The delay through the encoders is left unoptimized to illustrate the total combinational delay of the circuit; however, this delay can easily be hidden by pipelining the encoder in order to partition the combinational delay. The decoder implementation results are identical to the encoder results.

13.4.4.2 Xilinx 0.35 μm FPGA Process

The encoders were also synthesized, mapped, and implemented for a *Xilinx VirtexI-IPro*, Field Programmable Gate Array (FPGA) which used a 0.35 μm CMOS process. Figure 13.11 shows the Xilinx FPGA target that was used for implementation in addition to the test setup for the encoders.

Encoders for effective bus sizes of 2, 4, 6, and 8 were implemented using both the aggressive and non-aggressive constraints. Table 13.4 lists the delay and area impact of the bus expansion encoders when implemented in the FPGA process. In

Fig. 13.11 Xilinx FPGA target and test setup for encoder implementation

Table 13.4 Bus expansion encoder synthesis results in a 0.35 μm, FPGA process

	Bus size (m)	Noise limit
	—	5% (aggressive) & 10% (non-aggressive)
Delay (ns)	2	0.351
	4	1.020
	6	1.450
	8	1.610
FPGA Usage	2	<1%
	4	<1%
	6	<1%
	8	<1%
FPGA Implementation	2	3×, 2-Input FG's
	4	6×, 4-Input FG's
	6	9×, 6-Input FG's
	8	12×, 8-Input FG's

all cases the bus expansion encoder designs took less than 1% of the FPGA resources. The encoders were implemented using standard combinational logic blocks (Function Generators = FG's) within the FPGA, which resulted in minimal propagation delay through the circuit. Again, the unoptimized delay is presented in this table.

13.4.5 Measurement Results

The outputs of the FPGA were monitored using the 16950A Logic Analyzer from *Agilent Technologies, Inc.* The logic analysis measurements verified that the encoders could be taken from the concept to final implementation using standard IC design practices. Figures 13.12–13.15 show the logic analyzer measurement results for the aggressively constrained bus expansion encoders for effective sizes of 2, 4, 6, and 8.

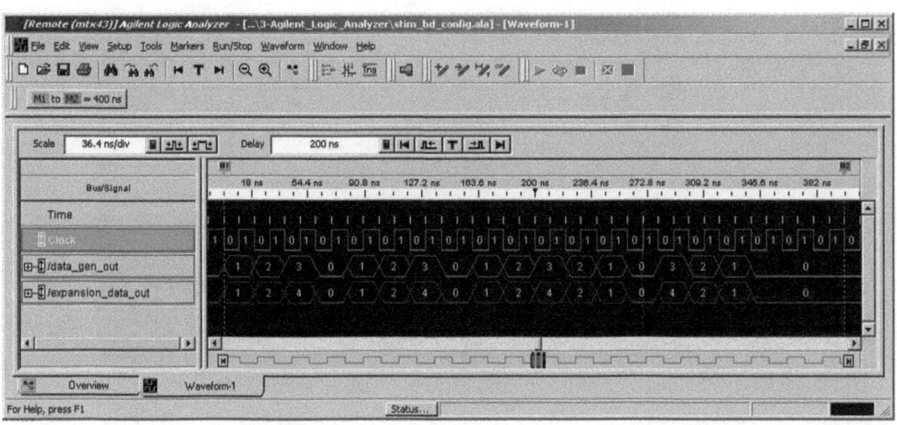

Fig. 13.12 Logic analyzer measurements of a 2-Bit bus expansion encoder

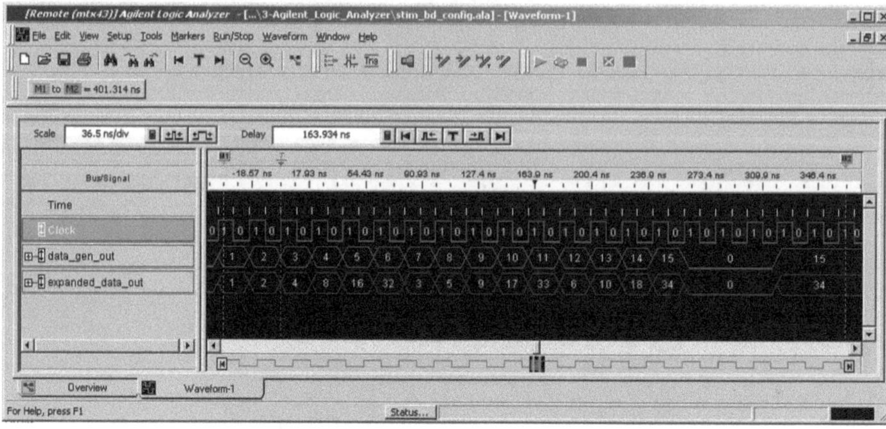

Fig. 13.13 Logic analyzer measurements of a 4-Bit bus expansion encoder

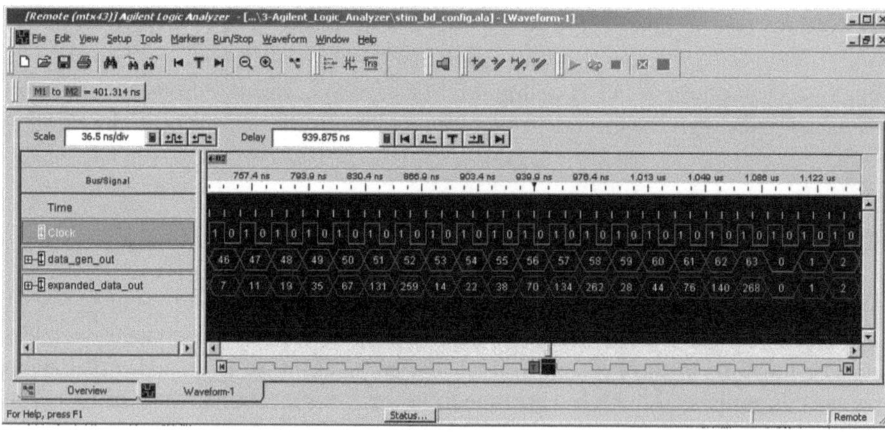

Fig. 13.14 Logic analyzer measurements of a 6-Bit bus expansion encoder

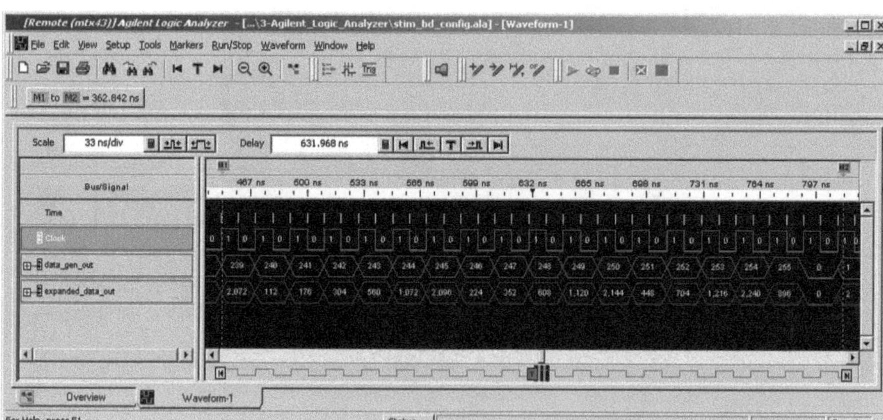

Fig. 13.15 Logic analyzer measurements of a 8-Bit bus expansion encoder

For each waveform, the m-bit clock and data are plotted in addition to the encoded n-bit data which is transmitted off-chip. The m-bit vectors are monitored using a debug port that routes the internal nodes of the on-chip bus to the logic analyzer. In practice, this debug port would be removed so not to cause unintentional SSN. These figures show that the actual implementation results of the encoder match the functional simulations. For the bus configuration of this FPGA (SPG $= 3:1:1$), the noise was reduced from 16% to 4% using the bus expansion encoding technique.

Chapter 14
Bus Stuttering Encoder

It was shown in Chap. 13 that the performance of an off-chip bus could be increased by avoiding a subset of patterns which resulted in noise greater than a specified limit. By eliminating the patterns that create the greatest amount of noise, the remaining patterns can be transmitted at a faster per-pin datarate without exceeding the user-defined noise limits of the system. The bus expansion encoding technique was successful in increasing the overall throughput of the bus even after considering the encoder overhead. The bus expansion encoder is ideal for applications where additional signal pins on the package are available for use when encoding the on-chip vectors; however, in the case where the number of package signal pins equals the on-chip bus width, then a different approach can be used.

In this chapter a *bus stuttering* encoder is presented. In the stutter encoder, if back-to-back vectors (or states) that are being transmitted off-chip induce a transition which results in a noise limit violation, an intermediate state is inserted between them. This intermediate or *stutter* state offers a method to transition between the two original states (without directly switching between them) in such a manner that each of the intermediate transitions do not result in a noise limit violation. It is possible that more than one stutter state is required in some cases. In this fashion, the transition between the original states is performed by inserting one or more stutter states. During the time when the stutter states are being transmitted off-chip, the clock that is used to latch the data on the receiver is gated out. This assumes a source synchronous data transmission scheme. This allows the receiver to be implemented using standard circuitry. The receiver will only acquire valid data since it only latches data when a valid clock edge is present. The number of stutter states that are used in the data transmission will depend on the aggressiveness of the user-defined noise limits.

14.1 Encoder Construction

The steps in creating the stutter encoder are similar to those used in constructing the bus expansion encoder up to the point where the directed graph is processed. The user still writes constraint equations which are evaluated for each possible transition on the

C. Duan et al., *On and Off-Chip Crosstalk Avoidance in VLSI Design*,
DOI 10.1007/978-1-4419-0947-3_14, © Springer Science+Business Media LLC 2010

bus. Transitions which result in a user-defined noise limit violation are removed from the subset of legal transitions. The remaining legal transitions are used to create the directed graph G which represents all of the legal paths between any two vertices. The graph is represented implicitly and efficiently using ROBDDs [20, 22]. At this point, the user can run either the bus expansion algorithm or the bus stuttering algorithm on the directed graph G.

14.1.1 Encoder Algorithm

The encoder is constructed by evaluating each vertex v_s of $G(V, E)$ and finding the shortest path between v_s and any destination vertex v_d in G using only legal edges of G (including self-edges). In order to be able to construct a stutter encoder, the following conditions must hold:

- There must exist at least two outgoing edges (including the self-edge) for each $v_s \in G$.
- There must exist at least two incoming edges (including the self-edge) for each $v_d \in G$.

These requirements ensure that for the directed graph G, each vertex can reach at least one other vertex and can be reached by at least one other vertex. If these requirements are not met, then the user must relax the user-defined noise limits until both conditions are met.

Given G, the algorithm first tests whether both of the above mentioned requirements are satisfied. The algorithm next attempts to determine the number of intermediate steps required for a vertex $v_s \in G$ to reach another vertex $v_d \in G$. If v_d can be reached with just one edge, the algorithm records the transition as a *direct* transition (one that requires 0 stutter steps).

For the case where v_s can *not* reach v_d with only one edge, then at least one stutter state is needed to complete the transition. The algorithm then attempts to find a path between v_s and v_d using two edges. Since the set of vertices V_d that can be reached from v_d by means of a direct path is known, the algorithm simply needs to find an edge from v_s to $v \in V_d$. Once such a path between v_s and v_d is found, then the algorithm records the intermediate vertex as a necessary *stutter state* which is required between v_s and v_d. This process is repeated for transitions which require more stutter states.

Algorithm 14.1 contains the pseudo-code for the stutter encoder algorithm. All *transition_path* variables are initially initialized to be empty. The routine *find_path*(v_s, v_d, l) returns a shortest path from the source v_s to destination v_d, with path_length l.

The maximum possible number of stutter states that may be used is $(2^{W_{bus}} - 1)$. This represents the worst-case where in order to transition from v_s to v_d, each and every other vertex within G must be used as a stutter state. While this represents the absolute theoretical worst case, experimental results have shown that the number of stutter states is typically between 0 and 3 for bus segments up to 8-bits. In real

Algorithm 14.1 Constructing the stutter encoder

for $(each\ v_s \in V)$ **do**
  for $(each\ v_c \in V)$ **do**
    for $(path\_length = 1;\ path\_length < 2^{W_{bus}-1};\ path\_length++)$ **do**
      if $(transition\_path[v_s, v_d]\ == \phi)$ **then**
        $transition\_path[v_s, v_d]\ = find\_path(v_s, v_d, path\_length)$
      end if
    end for
  end for
end for
$find\_edge(v_s, v_d, l)$
if $(\exists v_1, v_2, \cdots v_{(l-1)} S.T.((v_s, v_1) \in E) \wedge ((v_2, v_2) \in E) \wedge \cdots \wedge ((v_{(l-1)}, v_d) \in E))$
then
  $return\ (v_s, v_1, v_2, \cdots, v_{(l-1)}, v_d)$
else
  $return\ \phi$
end if

application of the stutter encoder, the actual number of stutter states that would ever be inserted within any data sequence is $W_{bus} - 1$ since this represents the total number of bits that could ever switch between two data vertices. It should be noted that this analysis is only performed on a representative bus segment (which is typically very small compared to the entire off-chip bus).

The construction of the encoder presented in this monograph is targeted at improving the performance of the off-chip data transmission. For this application, the shortest path between vertices is selected as the optimal choice for achieving maximum throughput. This encoder can be modified for optimization of other constraints such as synthesis complexity, circuit area, or circuit delay by selecting other routes between vertices other than the shortest path.

14.1.2 Encoder Overhead

To calculate the overhead of the stutter encoder, it is assumed that each vector $v_s \subseteq V$ has an equal probability of occurring on the bus. Using this assumption, a sequence of data patterns is constructed, in which each and every sequence occurs on the bus at least once. When this sequence is transmitted, the minimum number of stutter states are inserted in the transition between any pair of vectors. The maximum number of stutter states that will be inserted in a sequence for any given encoder is $2^{W_{bus}-1}$. Equation 14.1 gives the overhead of the stutter encoder.

$$Overhead_{bus\text{-}stuttering} = \left(\frac{\sum_{k=1}^{2^{(W_{bus}-1)}} (\#Transitions\ Requiring\ k\ Stutters) \cdot k}{2^{(2 \cdot W_{bus})}} \right) \cdot 100 \tag{14.1}$$

14.2 Decoder Construction

The stutter encoding technique assumes a source synchronous clocking architecture. In source synchronous clocking, the bus clock is generated at the transmitter and synchronized to the off-chip data being transmitted. The clock is then transmitted along with the data in the off-chip bus. By doing this, the timing correlation between the clock and data is extremely tight. This architecture has seen wide adoption in industry as a way to address channel-to-channel skew and common mode noise.

The stutter encoding technique is specifically designed for a source synchronous architecture. The encoding circuitry gates out the source synchronous clock when stutter states are transmitted off-chip. Since the receiving circuitry only aquires data on the rising edge of the source synchronous clock, the stutter states are ignored. In this manner, no special decoding circuitry is needed.

14.3 Experimental Results

To validate the feasibility of the stutter encoding technique, experimental results were performed on an example bus segment. For this example the off-chip bus was implemented using a BGA wire bonded package with parasitics given in Table 9.1. The off-chip bus uses a fixed $\frac{di}{dt}$ of 8 $\frac{MA}{s}$. Bus segments with between 2 to 8 signal pins were encoded using the stuttering technique, assuming that each segment had one V_{DD} pin and one V_{SS} pin. This bus was encoded using two sets of constraints – *aggressive* (P_{supply}, P_{gnd}, P_0, P_1, P_{-1}, $-P_{BW}$, and P_Γ set to 5%.) and *non-aggressive* (P_{supply}, P_{gnd}, P_0, P_1, P_{-1}, $-P_{BW}$, and P_Γ set to 10%). For these bus configurations and noise limits, all possible transitions were evaluated using the constraint equations described in Sect. 13.4. The remaining legal transitions were used in creating the directed graph G which was used to construct the stuttering encoder. Using Algorithm 12, the shortest paths between all possible vertices were found. In addition, the number of stutter states required to complete the intended transition were found. Table 14.1 lists the percentage of transitions that require stutter states for each of the bus sizes in this example. Using Eq. 14.1, the overhead of each of the encoders was calculated. Figure 14.1 shows the overhead of the stutter encoders as a function of bus segment size. Observe that for bus segment sizes less than 4-bits (aggressive constraints) and 6-bits (non-aggressive constraints), the overhead is less than 10%.

Figure 14.2 shows the bus throughput improvement when using the stutter encoding technique. This figure shows the percentage of throughput improvement of the bus (including the overhead of the encoder). The improvement of the *aggressive* encoder reaches 225% at a bus size of 6-bits. Beyond 6-bits, the overhead of the stutter encoder begins to outweigh the increased per-pin datarate (due to the increasing number of stutter states needed to avoid the noise limits).

Table 14.1 Percentage of transitions requiring stutter states

Noise limit	Bus size	Number of Stutter States			
—	—	0	1	2	3
5% Limit	2	100	0	0	0
	3	96.9	3.1	0	0
	4	89.8	10.2	0	0
	5	79.3	20.5	0.2	0
	6	66.6	32.5	0.9	0
	7	53.4	44	2.6	0
	8	41.1	53.4	5.4	0.1
10% Limit	2	100	0	0	0
	3	100	0	0	0
	4	99.2	0.8	0	0
	5	96.9	3.1	0	0
	6	92.5	7.5	0	0
	7	85.9	14.1	0	0
	8	77.3	22.6	0.1	0

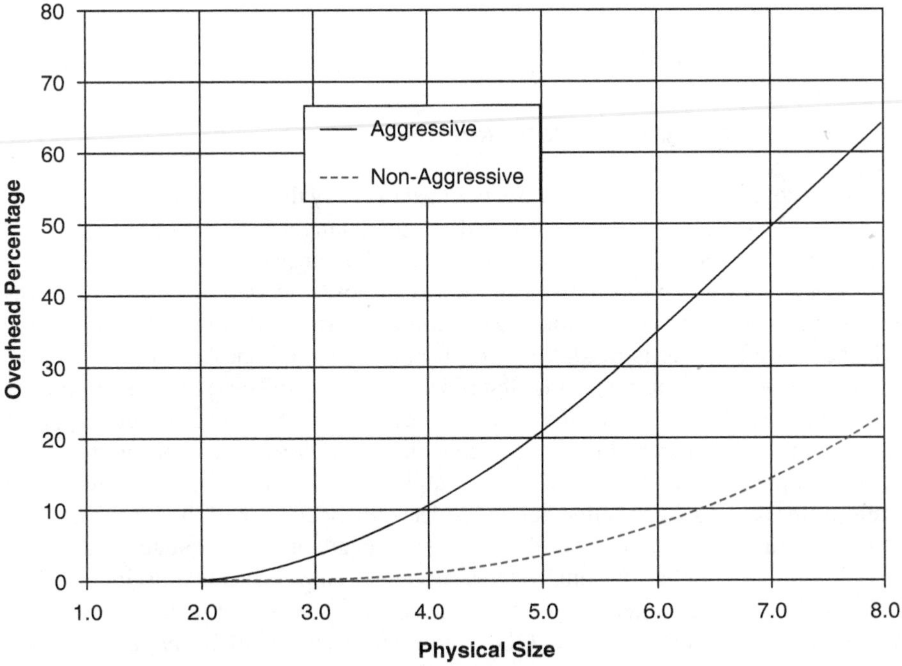

Fig. 14.1 Bus stuttering encoder overhead for the fixed $\frac{di}{dt}$ example

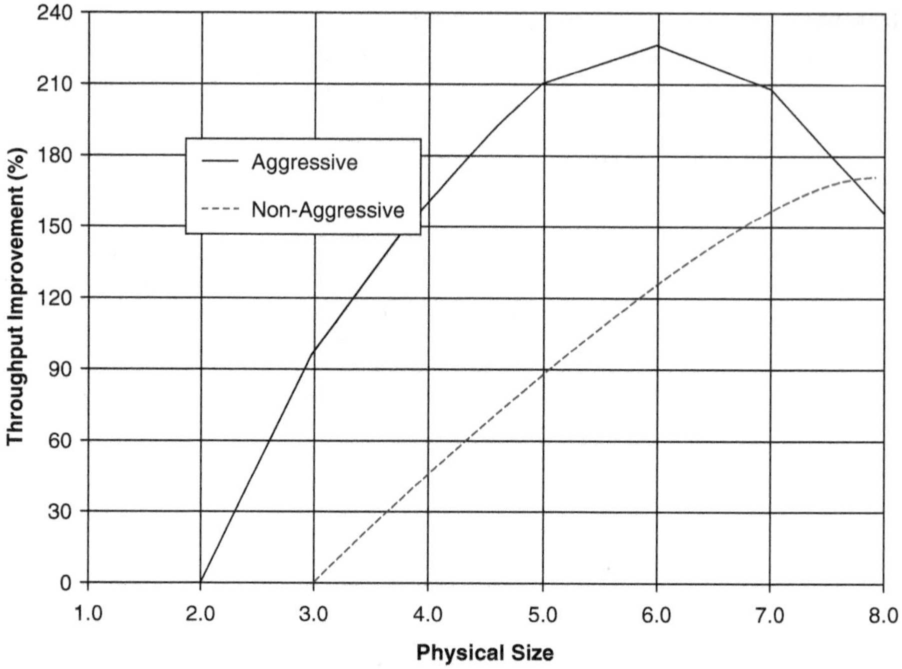

Fig. 14.2 Bus stuttering encoder throughput improvement

14.3.1 Functional Implementation

The bus stuttering encoders were implemented to verify their functionality and feasibility. Once again, the Verilog hardware description language was used to implement the encoder circuitry [29, 78]. The implementation consists of a pipeline in which each stage of the pipe is routed to a multiplexer whose output drives the bus patterns off-chip. A state machine monitors the incoming data from the core of the IC to check whether a stutter state is needed in the off-chip transmission. At the beginning of circuit operation, the output of the first pipeline stage is selected to be the output of the multiplexer. When a sequence of vectors occurs which require a stutter state, then the multiplexer is switched to the state machine input where the appropriate stutter state is output. After the stutter state(s) is output, then the state machine switches the multiplexer back to the pipeline but now selects the next stage of the pipe. The state machine continues to monitor the pipeline for illegal consecutive states and inserts the appropriate number of stutter states in the off-chip data transmission. During the time in which a stutter state is being selected as the output of the multiplexer, the control state machine gates out the source synchronous clock. By gating out the clock, it is ensured that the receiver will not latch any of the stutter states.

The pipeline in the stutter encoder is used to compare consecutive states within the data sequence. Each time that a stutter state is output on the bus, the encoded data sequence falls one clock period behind the original data. The depth of the pipeline is dependant on how many stutter states could potentially occur on the bus. The

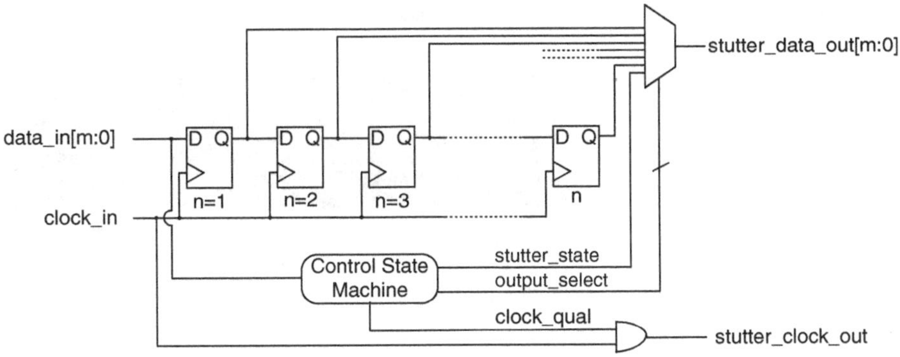

Fig. 14.3 Bus stuttering encoder schematic

bus protocol dictates the maximum length of consecutive data that could occur on the bus. This, in turn, dictates the maximum depth of the pipeline that is needed to prevent overflow. In practice, bus protocols have scheduled idle periods when the pipeline can be reset [54, 77, 82]. For this work, 32 pipeline stages were used in the implementation of the encoder. Figure 14.3 shows the schematic for the stutter encoder circuit.

Figures 14.4–14.6 show the Verilog simulation results for aggressively constrained bus stuttering encoders for sizes of 4, 6, and 8 bits. For each waveform the original data and clock are plotted in addition to the encoded data and gated clock which are transmitted off-chip.

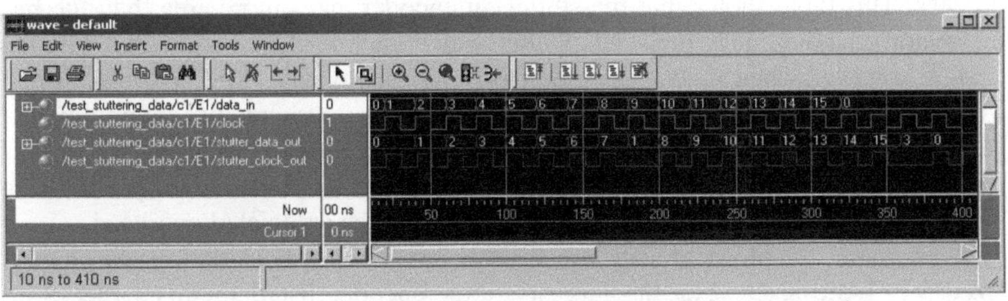

Fig. 14.4 Verilog simulation results for a 4-Bit bus stuttering encoder

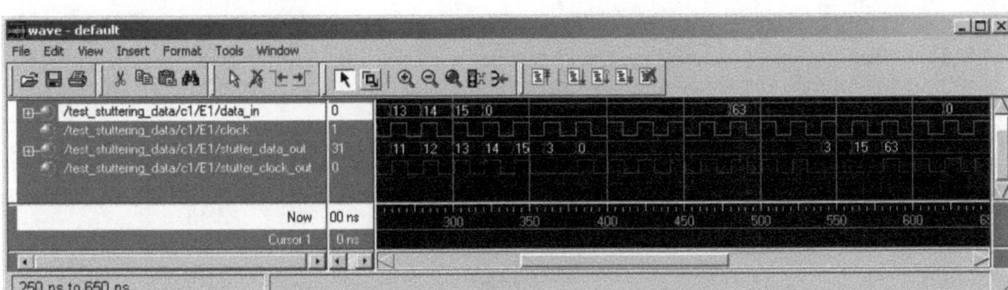

Fig. 14.5 Verilog simulation results for a 6-Bit bus stuttering encoder

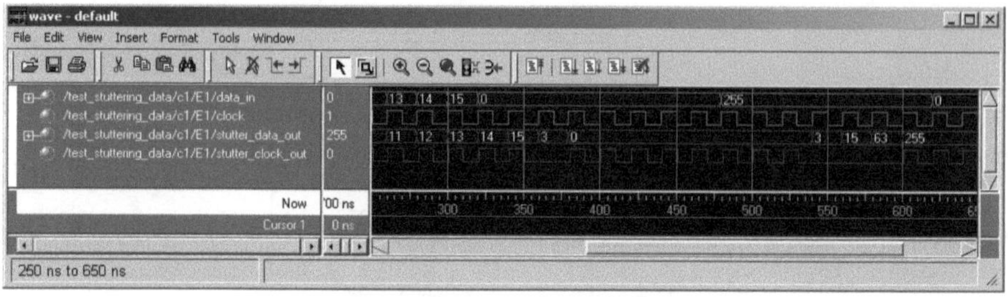

Fig. 14.6 Verilog simulation results for a 8-Bit bus stuttering encoder

14.3.2 Physical Implementation

In order to validate the feasibility of the stutter encoder, the physical design of the CODEC was performed.

14.3.2.1 TSMC 0.13 µm ASIC Process

The stutter encoders were synthesized using the TSMC 0.13 µm CMOS IC process to quantify their on-chip delay and area. Encoders for bus sizes of 4, 6, and 8 were implemented. For each of these sizes, both the aggressive (5% of V_{DD}) and the non-aggressive (10% of V_{DD}) stutter encoders were synthesized. Table 14.2 lists the delay and area impact of the bus stuttering encoders when implemented in a TSMC 0.13 µm process. This table shows that the stuttering encoder takes more area than the bus expansion encoder due to the state machines and pipeline stages associated with the design; however, when analyzing modern off-chip busses such as PCI Express, Rapid I/O, and HyperTransport, it is found that all of these busses already contain pipeline stages and state machines to handle encoding/decoding required by the protocol. This means that the stuttering encoder design could simply be included in the pre-existing state machine and pipeline stages, thereby reducing the associated overhead.

In any case, the physical sizes are still negligible when considering reasonably-sized VLSI die sizes. For the largest stutter encoder listed in this table (8-bit aggressively encoded), the area consumed is still less than 1.5% of a 5 mm^2 die. The un-optimized circuit delay is reported, which could be reduced using more

Table 14.2 Bus stuttering encoder synthesis results in a TSMC 0.13 µm process

	Bus size	Noise limit	
	—	5% (aggressive)	10% (non-aggressive)
Delay (ns)	4	2.02	1.99
	6	2.42	2.38
	8	2.85	2.79
Area (um^2)	4	311k	310k
	6	362k	345k
	8	382k	368k

Table 14.3 Bus stuttering encoder synthesis results for a Xilinx VirtexIIPro FPGA

	Bus size	Noise limit
	—	5% (aggressive) & 10% (non-aggressive)
Delay (ns)	4	4.78
	6	5.29
	8	5.89
FPGA Usage	4	<1%
	6	<1%
	8	<1.5%

advanced circuit techniques. For the stutter encoder circuit, the combinational logic in the control state machine was the largest source of delay in the system.

14.3.2.2 Xilinx 0.35 μm FPGA Process

The encoders were also synthesized, mapped, and implemented for a *Xilinx VirtexIIPro*, Field Programmable Gate Array (FPGA). Stutter encoders for bus sizes of 4, 6, and 8-bits were implemented using both the aggressive and non-aggressive constraints. Table 14.3 lists the delay and area impact of the bus stuttering encoders when implemented in the FPGA process. In all cases, the bus stuttering encoder designs took up less than 1.5% of the total FPGA resources.

14.3.3 Measurement Results

Once again, the test setup in Figure 13.11 was used to verify the actual performance of the encoders. The outputs of the FPGA were again monitored using the 16950A Logic Analyzer from *Agilent Technologies, Inc.* Figures 14.7–14.9 show the logic analyzer measurement results for aggressively constrained bus stuttering encoders

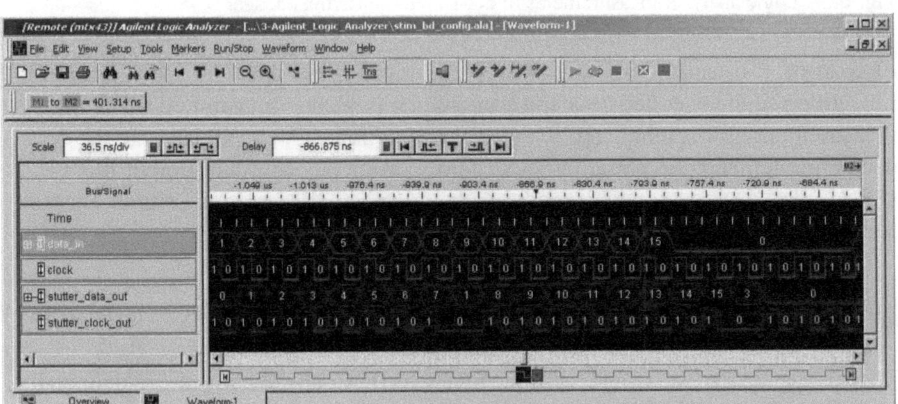

Fig. 14.7 Logic analyzer measurements of a 4-Bit bus stuttering encoder

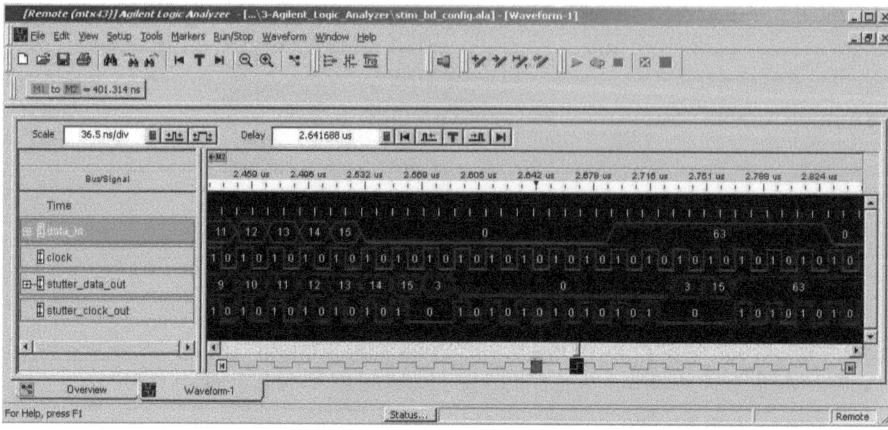

Fig. 14.8 Logic analyzer measurements of a 6-Bit bus stuttering encoder

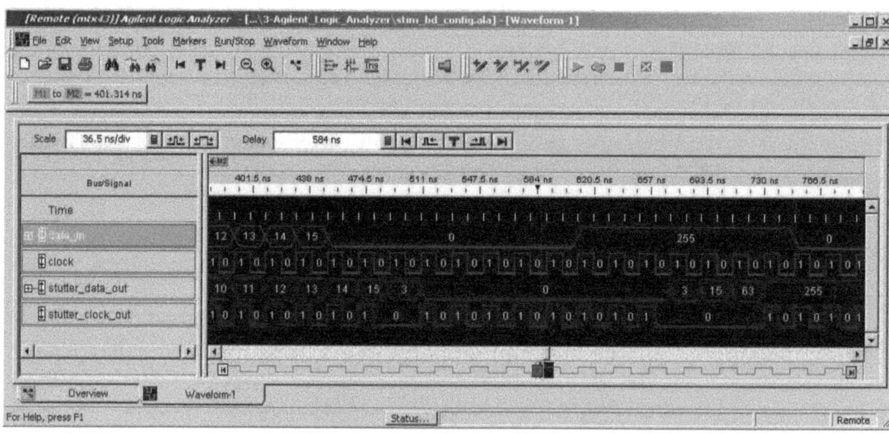

Fig. 14.9 Logic analyzer measurements of a 8-Bit bus stuttering encoder

of sizes of 4, 6, and 8. For each waveform, the original data and clock are plotted in addition to the encoded data and gated clock (which is transmitted off-chip). The measurement results match the functional simulations which validate that the design can be implemented in a real target system. For the bus configuration of this FPGA (SPG $= 3:1:1$), the noise was reduced from 16% to 4% using the stutter encoding technique.

14.3.4 Discussion

The bus expansion encoding technique presented in Chap. 13 and the stutter encoding technique presented in this chapter both avoid patterns on the off-chip bus which result in noise limit violations. Both encoders are shown to improve performance even after considering the overhead of the encoder. When to use one encoding scheme versus the other depends on the application. The bus expansion encoder is aimed at generic off-chip busses which do not contain any special data processing circuitry for the off-chip transmission. In these cases, the encoder area impact is minimal since it is the only circuitry placed in the path of the off-chip data.

The bus stuttering encoder is targeted at source synchronous, protocol based busses. In these cases, the pipeline and control state machines already exist to handle the protocol of the bus. Adding stutter encoding capability to the preexisting circuitry is incremental and can potentially have less area impact than would the bus expansion encoder depending on how aggressively the noise is constrained.

To select the appropriate encoding technique, the designer must analyze the area impact on the design. For wider, parallel busses the bus expansion encoder is the better choice. For source synchronous, protocol based busses the bus stuttering encoder is the better choice.

Chapter 15
Impedance Compensation

Impedance mismatches between the IC package and the system PCB cause reflections that lead to unwanted noise in the system. This noise limits the maximum datarate that a package can achieve (Eq. 11.26). As described in Sect. 8.2, the impedance of the package interconnect is dictated by the self-inductance and self-capacitance of the physical structure in addition to its mutual inductance and mutual capacitance with respect to neighboring structures. Since the package interconnect was originally developed for mechanical robustness, the electrical parasitics of the interconnect structure can be considerable. In the majority of cases the resulting impedance of the package interconnect is not matched to that of the system. In addition, the mechanical structures are difficult to alter due to the complex manufacturing processes required to construct the interconnect. This makes it difficult and expensive to change the physical structure of the package to match the characteristic impedance to that of the system PCB.

Chapters 13 and 14 presented bus encoding techniques that can reduce the effective mutual inductive and mutual capacitive contributions to the impedance of a given signal pin. This is accomplished by avoiding the switching patterns which induce noise of a sufficiently high magnitude. This reduces the worst-case mutual inductive and mutual capacitive coupling the bus. By reducing the coupling contribution for a given signal pin, the variation in the impedance of the structure is reduced, thereby reducing the magnitude of the worst-case reflections from the package. While these encoding techniques address the mutual coupling between signal pins in the package, they do not address the self-inductance and self-capacitance of the package structure. As illustrated in Tables 9.1 and 9.2, the self-inductance and self-capacitance of the interconnect structure contribute the majority of the electrical parasitics to the characteristic impedance of the signal pin. Also illustrated in these tables is that the package interconnect is predominately inductive. The inductive nature of the interconnect leads to a relatively high impedance when compared to the majority of impedances used in system PCBs, which typically range between 50 to 75 Ω. The higher impedance of the package pins cause positive reflections according to Eq. 8.13.

To reduce the impedance discontinuities introduced by the electrical parasitics of the package interconnect, this chapter introduces an *impedance compensation* technique. In this technique, additional *compensation* capacitance is added near the

C. Duan et al., *On and Off-Chip Crosstalk Avoidance in VLSI Design*,
DOI 10.1007/978-1-4419-0947-3_15,

package interconnect in order to reduce the characteristic impedance of the structure. By inserting the compensation capacitance near the inductive interconnect, the overall impedance can be reduced by increasing the net capacitance in Eq. 8.11. Two styles of are presented, *static* and *dynamic*. In the static compensator, predefined capacitance is placed on the package and on the IC to *surround* the level 1 interconnect. In the dynamic compensation approach, programmable capacitors are placed on-chip that can be controlled by software to achieve the best impedance match. For both compensators, two styles of on-chip capacitors are examined. The first on-chip capacitor is a *Metal-Insulator-Metal* (MIM) device. The second is a *Device-Based* capacitor. Each type of capacitor is evaluated for variability across bias voltage values, and for implementation area in the context of impedance matching of the package structure.

15.1 Static Compensator

This section describes the technique to match the characteristic impedance of the package interconnect to that of the system PCB.

15.1.1 Methodology

Tables 9.1 and 9.2 illustrate that the interconnect for the packages studied in this work are mostly *inductive* when considering impedance. These tables also illustrate that the level 1 interconnect is the largest source of excess inductance within the package. Specifically, the wire bond interconnect is the largest contributor of inductance within any of the packages. Further, it is assumed that if the compensator is able to address the impedance associated with the wire-bond, it will also be able to address the inductance of any other package structure.

In the static compensator methodology, a predefined capacitance is placed on both sides of the level 1 interconnect. When considering the wire-bond interconnect, the compensation capacitance is placed on-chip as well as on-package. The on-chip capacitance (C_{comp1}) is placed near or beneath the ball bond pads of the wire-bond. The on-package capacitance (C_{comp2}) is placed near or beneath the wedge bond pads of the wire-bond. Figure 15.1 shows the optimal locations of the static compensation capacitance in a wire-bonded system.

The static compensation capacitors will alter the characteristic impedance of the package by increasing the capacitance value in Eq. 8.11. Equations 15.1 and 15.2 provide the total compensation capacitance and characteristic impedance of the wire-bond interconnect after applying the static compensation capacitance.

$$C_{comp} = C_{comp1} + C_{comp2} \tag{15.1}$$

$$Z_{0-static} = \sqrt{\frac{L_{wb}}{C_{wb} + C_{comp1} + C_{comp2}}} \tag{15.2}$$

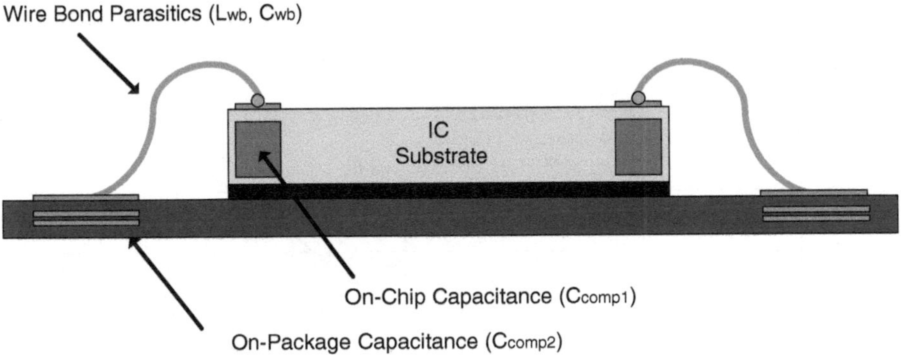

Fig. 15.1 Cross-section of wire-bonded system with compensation locations

15.1.2 Compensator Proximity

In order to model the level 1 inductance and compensation capacitance as a single lumped element, the compensation capacitance must reside within a certain distance of the inductance. The maximum distance of the compensation capacitance from the interconnect inductance depends on the frequency components present in the forcing function of the incident digital signal. If the compensation capacitance resides spatially close to the inductance, the resulting structure can be modeled as a lumped element, and the impedance of the structure is then deterministic. Typically, a structure is considered a lumped element when its electrical length (i.e. its propagation delay) is less than 20% of the highest risetime in the system [57]. Figure 15.2 shows the physical length at which a structure can be treated as either a lumped or distributed element as a function of risetime (for several common package materials). Given a particular risetime, physical lengths less than indicated in this plot are treated as a lumped element. For physical lengths greater than this amount, the structure is treated as a distributed element. This plot considers the propagation delay of four common dielectric materials that are used in VLSI packaging.

For VLSI risetimes less than 1ns, this plot shows that structures with electrical lengths less than 200 ps can be treated as lumped elements. Since standard package substrates utilize dielectric constants between Dk = 3 to Dk = 5, physical lengths less than approximately 1 inch may be considered lumped elements. This means that implementing embedded capacitors within the package or IC substrate can directly alter the characteristic impedance of the package interconnect (using a lumped model to compute the altered characteristic impedance). Specifically, embedding capacitors on-chip or on-package will be utilized to alter the characteristic impedance of the package interconnect.

15.1.3 On-Chip Capacitors

Two styles of on-chip capacitors were evaluated for use to implement C_{comp1} in the static compensator design. The first style of on-chip capacitor is called a

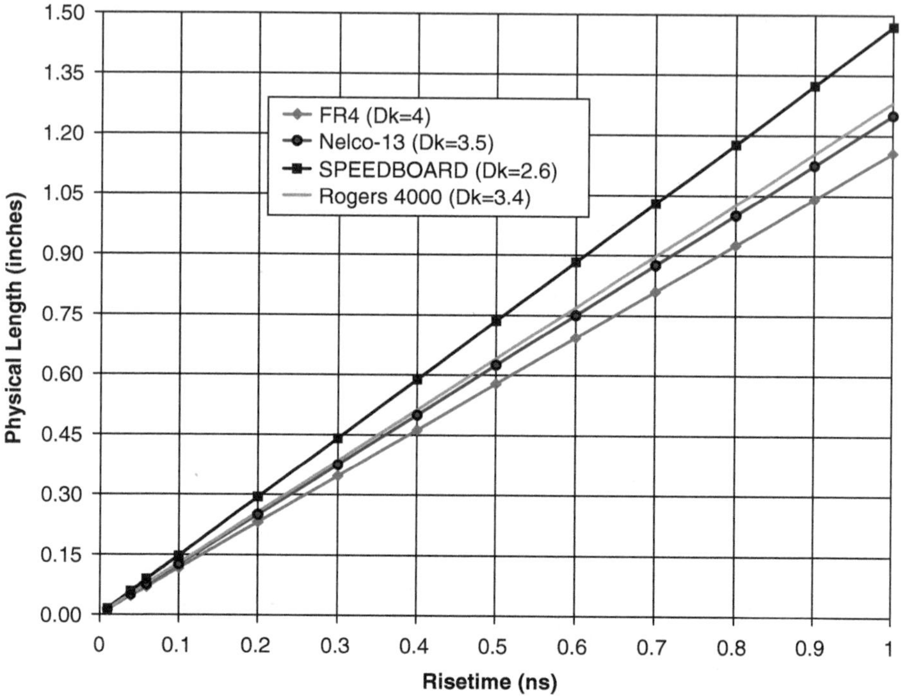

Fig. 15.2 Physical length at which structures become distributed elements

Device-based capacitor. This capacitor is created using the same process technology utilized in creating an NMOS transistor. The lower conducting plate of the capacitor is the heavily doped, p-type channel of the silicon transistor. The insulating material is the gate oxide of the NMOS transistor, which is a thin layer of silicon-oxide (SiO_2) above the channel of the device. The upper plate of the capacitor is the *PolySilicon* (*poly-Si*) gate of the NMOS device. The drain and source terminals are connected together and form one electrode of the capacitor, while the PolySilicon gate forms the other electrode. This style of capacitor has a very high capacitance density due to the thin plate separation which is identical to the gate thickness (t_{ox}) of the device. In this work, a 0.1 μm, BPTM process [1] is used to implement the on-chip capacitors. Using this technology, oxide layers as thin as 25 Å can be realized with a dielectric constant of $D_k = 3.9$. This translates into a capacitance density of 13.7 fF/um^2.

While the device-based capacitor achieves the highest density of any other style of on-chip capacitor, it suffers from two major drawbacks. The first drawback is that the capacitance value will change as a function of bias voltage. This nonlinearity is due to the fact that as bias voltage changes, the device goes from cut-off to weak inversion and strong inversion. The nonlinearity of the capacitance as a function of bias voltage is a problem in digital applications, since the bias voltage typically varies from V_{SS} to V_{DD}. The second drawback of this style of capacitor is that it consumes valuable area within the IC that is normally reserved for the implementation of digital circuitry.

The second type of on-chip capacitor studied in this work is called a *Metal-Insulator-Metal* (MIM) capacitor. In a MIM capacitor, an additional process step is used to form an extra metal layer in the upper layers of the IC (typically above metal 3). This extra process step inserts an additional metal layer above the standard metal 3 layer such that the insulator thickness between this new layer and metal 3 is much lower than the inter-layer dielectric between metal 3 and 4. This extra step allows the creation of a parallel plate capacitor that uses a standard metal layer as the lower plate (metal 3) and an additional MIM layer as the upper plate (metal 4'). Since the extra MIM layer is located spatially closer to the lower plate, a higher capacitance density is achieved compared to the capacitor with metal 3 and metal 4 plates. In addition, the extra MIM layer does not take up any routing area on metal 4. MIM technology is able to achieve plate-to-plate separations as thin as $t_{ox} = 100$ Å, which translates into a capacitance density of 1.15 fF/um^2. While MIM capacitors have a lower capacitance density relative to device-based capacitors, their capacitance is constant over the full range of bias voltages. This is because both plates within the MIM capacitor are formed with standard metal with passive materials between the plates.

For the static compensator, both styles of on-chip capacitors are investigated. For each capacitor, the nonlinearity with bias voltage and area utilization are quantified. Figure 15.3 shows the cross-section for both styles of on-chip capacitors.

15.1.4 On-Package Capacitors

Embedded PCB capacitors are evaluated to implement C_{comp2} in the static compensator design. An embedded PCB capacitor (EC) is simply a parallel plate capacitor that is formed using standard metal layers within the PCB lamination. Embedded capacitors are preferred over surface mount (SMT) components due to the simplicity of their implementation. Since the capacitors are constructed using simple metal-to-metal area, no surface mount pads, loading process, or through-hole vias are needed. While ECs are easy to implement, they typically are only used for small values of capacitance (<10 pF). Standard PCB lamination processes are able to achieve plane-to-plane separation as low as 0.051 mm. Using a standard package dielectric such as FR4 ($D_k = 4$), this translates into a capacitance density of 420 fF/mm^2. Table 15.1 lists the capacitance densities and variation with bias voltages for all of the capacitors used in the compensator design. For each capacitor, a reasonable size of capacitor is used for illistrative purposes.

15.1.5 Static Compensator Design

The static compensator is designed to alter the impedance of the largest inductive component of the package interconnect (i.e., the wire-bond). It is assumed that if this design methodology is successful in compensating for the wire-bond inductance, it will also be successful in compensating for interconnect structures with less inductive parasitics (i.e., flip-chip bumps).

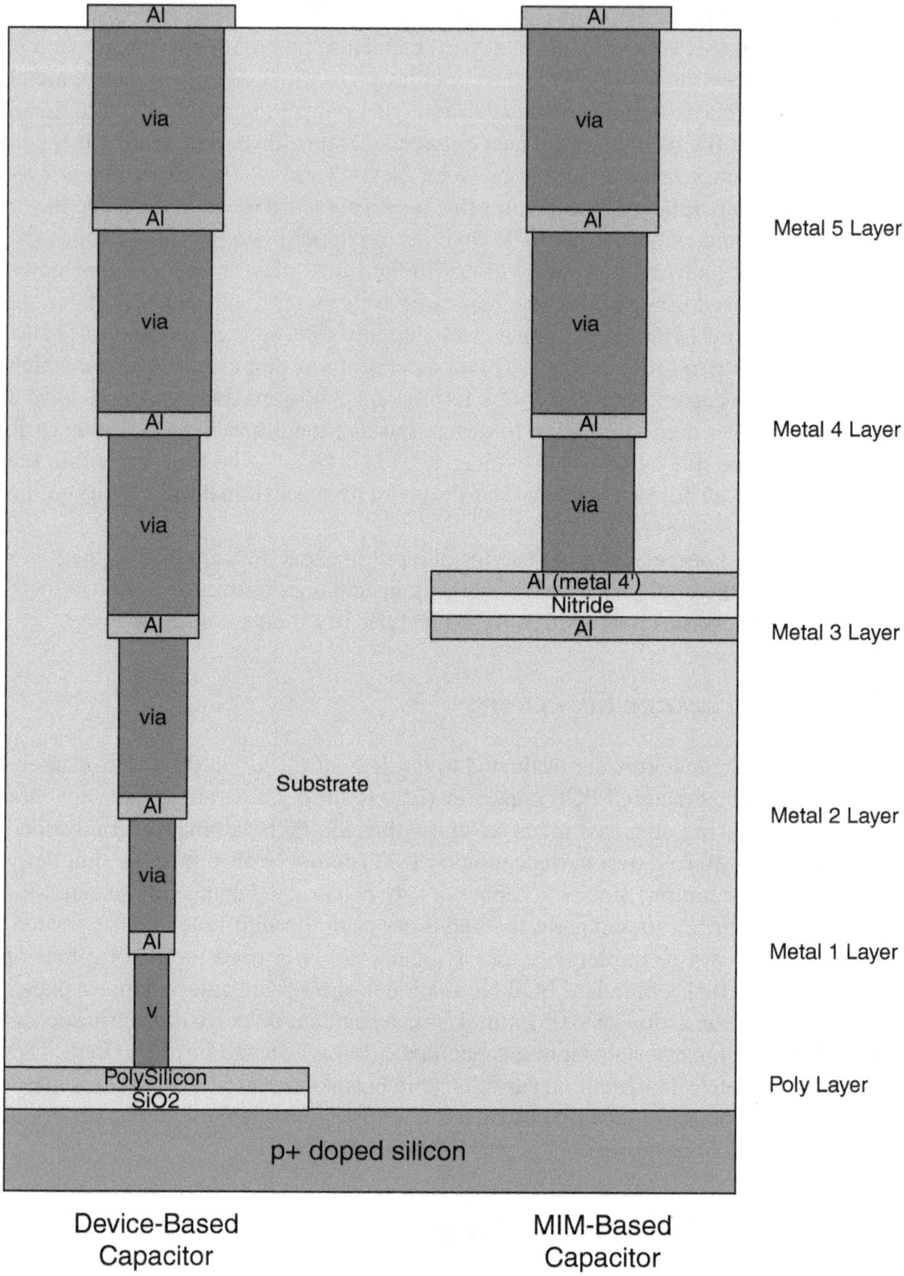

Fig. 15.3 On-chip capacitor cross-section

Table 15.1 Density and linearity of capacitors used for compensation

Structure	$C_{density}$	Size	C_{total}	Notes
MIM-Based Capacitor	1.15 fF/um^2	100 um^2	11.5 pF	$t_{ox} = 600$ Å$(Si_3N_4), D_k = 7$
Device-Based Capacitor ($V_{GS} = 0v$)	2.7 fF/um^2	100 um^2	27 pF	$t_{ox} = 25$ Å$(SiO_2), D_k = 3.9$
Device-Based Capacitor ($V_{GS} = 1.5v$)	13.8 fF/um^2	100 um^2	138 pF	$t_{ox} = 25$ Å$(SiO_2), D_k = 3.9$
Embedded Capacitor	420 fF/mm^2	0.5 mm^2	0.105 pF	$t_{ox} = 0.051$ mm$(FR4), D_k = 4$

Using the electrical parameters from Table 9.2 and solving Eqs. 15.1 and 15.2, the values of C_{comp1} and C_{comp2} for the static compensator can be found (assuming $C_{comp1} = C_{comp2}$). Table 15.2 lists the optimal capacitor values required to match the wire bond impedances (Table 9.2) to 50 Ω. Table 15.3 lists the corresponding sizes of the different types of capacitors required to realize C_{comp1} and C_{comp2}. This table lists the sizes for both the MIM-based and Device-based on-chip capacitor implementations of C_{comp1}.

15.1.6 Experimental Results

In order to evaluate the performance of the static compensator, SPICE simulations were performed on all lengths of wire bonds listed in Table 15.2. Figure 15.4 shows the simulated Time Domain Reflectrometry (TDR) of the static compensator (Eq. 8.14). A TDR simulation shows how much of a reflection (Γ) is caused by the wire bond. For each length of wire bond (1 to 5 mm), a 117 ps (3 GHz) input step is used to stimulate the wire bond. In Fig. 15.4, the TDR waveforms are offset in the voltage axis for view-ability with the 1 mm curve on the top and the 5 mm curve

Table 15.2 Static compensation capacitor values

$Length_{wb}$(mm)	C_{comp1}(fF)	C_{comp2}(fF)
1	102	102
2	208	208
3	325	325
4	450	450
5	575	575

Table 15.3 Static compensation capacitor sizes

$Length_{wb}$(mm)	$C_{comp1-MIM}$(µm)	$C_{comp1-Device}$(µm)	$C_{comp2-EC}$(µm)
1	10 × 10	2.7 × 2.7	388 × 388
2	14 × 14	3.9 × 3.9	554 × 554
3	18 × 18	4.9 × 4.9	692 × 692
4	21 × 21	5.8 × 5.8	815 × 815
5	24 × 24	6.5 × 6.5	921 × 921

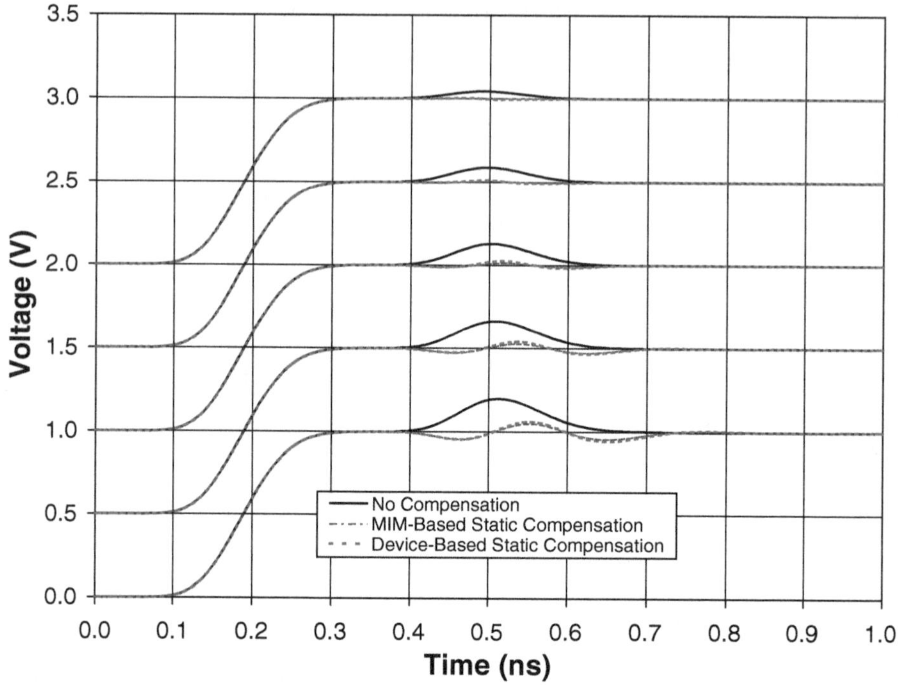

Fig. 15.4 Static compensator TDR simulation results

on bottom. Reflections off of an *inductive* interconnect (i.e., $Z_L \geq Z_0$) will result in the waveform traveling above its steady state value. Reflections off of a *capacitive* interconnect (i.e., $Z_L \leq Z_0$) will result in the waveform traveling below its steady state value. For each length of wire bond the non-compensated, MIM-based, and Device-based static compensation curves are shown. This figure shows the dramatic reduction in wire bond reflections when using a static compensator. Table 15.4 reports the reduction in reflections when using the static compensator(s).

Another way to observe the effect of the compensator is to observe the input impedance of the structure in the frequency domain. Figure 15.5 shows the input impedance of the wire bond structure versus frequency for the 3 mm wire bond. In this figure, the non-compensated, MIM-based, and Device-based static compensation curves are again shown. To compare the performance we record the frequency at which the input impedance deviates by 10 Ω from the target impedance. In this case,

Table 15.4 Reflection reduction due to static compensator

$Length_{wb}$(mm)	$\Gamma_{No-Comp}(\%)$	$\Gamma_{MIM-Comp}(\%)$	$\Gamma_{Device-Comp}(\%)$
1	4.5	0.05	0.5
2	8.7	0.4	1.2
3	12.7	1.3	2.4
4	16.4	2.7	4.1
5	19.8	4.8	6.0

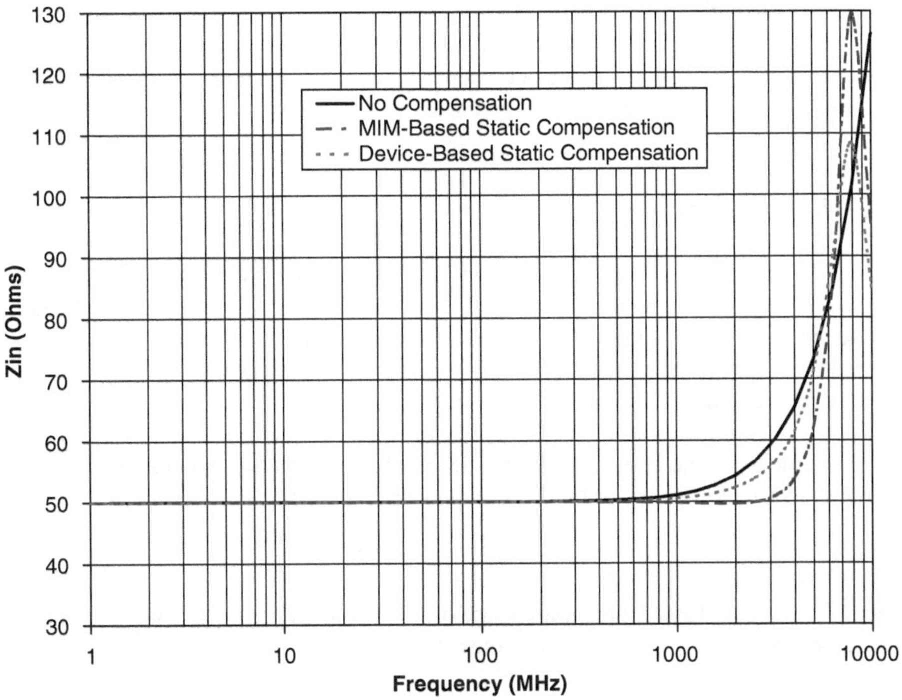

Fig. 15.5 Static compensator input impedance simulation results

the compensators are designed to match the structure to 50 Ω. This figure illustrates that adding a static compensator can keep the lumped impedance of the wire bond closer to 50 Ω up to a much higher frequency. For this example, the 3 mm wire bond was kept to within 10 Ω of the target impedance (50 Ω) up to 4.8 GHz when using the MIM-based compensator (compared to only 3.1 GHz when considering the uncompensated wire bond).

Table 15.5 lists the frequencies at which the input impedance strays to +/−10 Ω from the target for all of the lengths of wire bonds evaluated.

These results show the dramatic reduction in impedance discontinuities when using a static compensator. In all cases, the MIM-based compensator outperformed the Device-based compensator.

Table 15.5 Frequency at which static compensator is +/−10 Ω from design

$Length_{wb}$(mm)	$f_{No-Comp}$(GHz)	$f_{MIM-Comp}$(GHz)	$f_{Device-Comp}$(GHz)
1	9.3	14	12
2	4.7	7.1	5.7
3	3.1	4.8	3.8
4	2.4	3.7	2.9
5	1.9	3.0	2.5

15.2 Dynamic Compensator

This section describes the technique used to match the impedance of the package interconnect to the characteristic impedance of the system.

15.2.1 Methodology

In the dynamic compensator methodology, capacitance is placed only on-chip (C_{comp1}). The term *dynamic* means that the capacitance is programmable through active circuitry on the chip. The programmability of the compensation capacitance allows the compensator to successfully match impedance across variations in the wire bond inductance. The programmability implies that the designer does not need to know the exact wire bond inductance prior to IC fabrication. This has the advantage that the dynamic compensator can accommodate process and design variation. In the dynamic compensator the net compensation capacitance is given by:

$$C_{comp} = C_{comp1} \tag{15.3}$$

Using Eq. 8.11 to describe the impedance of the dynamically compensated structure, the impedance of the wire bond becomes:

$$Z_{0-dynamic} = \sqrt{\frac{L_{wb}}{C_{wb} + C_{comp1}}} \tag{15.4}$$

15.2.2 Dynamic Compensator Design

The design of the dynamic compensator consists of CMOS pass gates that connect to integrated binary-weighted, on-chip capacitors. In this design there are three integrated capacitors (C_1, C_2 and C_3) that can be switched in. The number of capacitors can be increased, however experimental results indicate that sufficient resolution can be achieved using just three capacitors. Each of these capacitors uses a pass gate to connect to the wire bond (Pass Gate #1, Pass Gate #2, Pass Gate #3). Each pass gate has a control signal which either connects/isolates the capacitors to/from the on-chip I/O pad, which is connected to one end of the wire bond. Figure 15.6 shows the schematic of the dynamic compensator design.

15.2.2.1 Capacitor Design

The diffusion regions associated with the pass gates contribute an additional capacitance to C_{comp1}, which must be considered in the design. For each of the three programmable capacitor banks, the net capacitance will be the sum of the diffusion

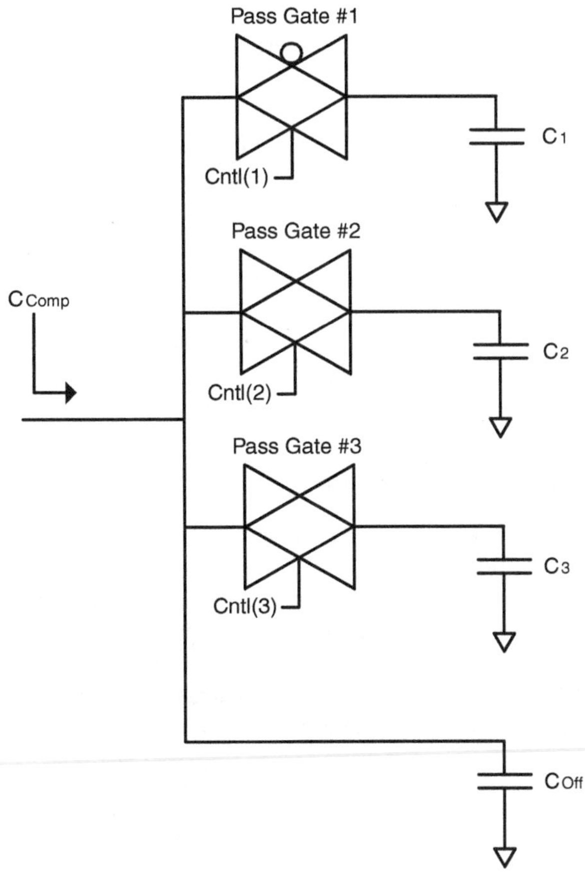

Fig. 15.6 Dynamic compensator circuit

capacitance of the pass gate and the integrated capacitor (Eqs. 15.5–15.7). If the pass gate of any bank i is turned off, then bank i simply contributes a capacitance C_{pgi} to C_{comp1}.

$$C_{Bank1} = C_{pg1} + C_1 \tag{15.5}$$

$$C_{Bank2} = C_{pg2} + C_2 \tag{15.6}$$

$$C_{Bank3} = C_{pg3} + C_3 \tag{15.7}$$

To achieve a larger programming range, the integrated capacitors are binary-weighted such that:

$$C_{Bank3} = 2 \cdot C_{Bank2} = 4 \cdot C_{Bank1} \tag{15.8}$$

Using three control bits, 8 programmable values of capacitance can be connected to the wire bond in increments of C_{Bank1}. C_{Off} is a constant, nonprogrammable capacitor which serves as an offset capacitance. It therefore sets the minimum capacitance value of the compensator. Using the conventions just described, the range of the compensator can be described as:

$$C_{min} = C_{Off} = C_{pg1} + C_{pg2} + C_{pg3} \tag{15.9}$$

$$C_{max} = C_{Off} + C_{Bank1} + C_{Bank2} + C_{Bank3} \tag{15.10}$$

$$C_{step} = C_{Bank1} \tag{15.11}$$

In this work the pass gates are implemented using a 0.1 μm CMOS process from BPTM. As with the static compensator, the integrated capacitors (C_1,C_2, C_3, C_{Off}) are implemented using the two different on-chip capacitance realization techniques that are evaluated in this work (MIM-based and Device-Based). Both of these capacitor techniques are evaluated for range, area efficiency, and nonlinearity for application in the dynamic compensator design.

15.2.2.2 Pass Gate Design

As described in the previous section, the diffusion capacitance of the pass gates will contribute to the total capacitance of each bank. The pass gate must be designed to have sufficient strength to drive the integrated capacitors (C_1, C_2, C_3). Typical CMOS design rules dictate that the pass gate sizing should be 1/3 of the size of an equivalent inverter that represents the integrated capacitor being driven [59]. This rule defines the amount of capacitance that will be present in each bank due to the pass gate and to the integrated capacitor.

$$C_{pg} = \frac{1}{3} \cdot C_{Bank} \tag{15.12}$$

$$C_{int} = \frac{2}{3} \cdot C_{Bank} \tag{15.13}$$

Using the Eqs. 15.5–15.13 and the values from Table 9.2, the sizing of the resulting compensation circuitry can be determined. Table 15.6 lists the values of the capacitors needed in the dynamic compensator that will match the impedance of the wire bond to 50 Ω (Eq. 15.4).

Table 15.7 lists the device sizes of the dynamic compensator circuit. The capacitances C_1, C_2 and C_3 are implemented as square devices for minimal area utilization. The total area of the compensator is computed as the size of the smallest enclosing square on the die. It is clear that the MIM-based dynamic compensator occupies more area than the Device-based compensator; however, the nonlinearity of both

Table 15.6 Dynamic compensation capacitor values

$Length_{wb}$(mm)	C_{comp1}(fF)
1	202
2	403
3	605
4	806
5	1008

Table 15.7 Dynamic compensation capacitor sizes

Component	MIM-Based Area (W × L)	Device-Based Area (W × L)
Pass Gate #1	32.4 × 0.1 μm	32.4 × 0.1 μm
Pass Gate #2	62.5 × 0.1 μm	62.5 × 0.1 μm
Pass Gate #3	129.6 × 0.1 μm	129.6 × 0.1 μm
C_{off}	8.5 × 8.5 μm	2.5 × 2.5 μm
C_1	11 × 11 μm	3.3 × 3.3 μm
C_2	15.5 × 15.5 μm	4.6 × 4.6 μm
C_3	22 × 22 μm	6.6 × 6.6 μm
Total	65 × 65 μm	25 × 25 μm

compensators must be analyzed to compare the applicability of the two circuits. In the next section, experimental results are presented that compare the two compensator designs.

15.2.3 Experimental Results

The same set of SPICE simulations as in the static compensator were performed on the dynamic compensator to evaluate its performance.

Figure 15.7 shows the simulated TDR of the dynamic compensator (Eq. 8.14). Each length of wire bond (1 to 5 mm) is evaluated when stimulated with a 117 ps

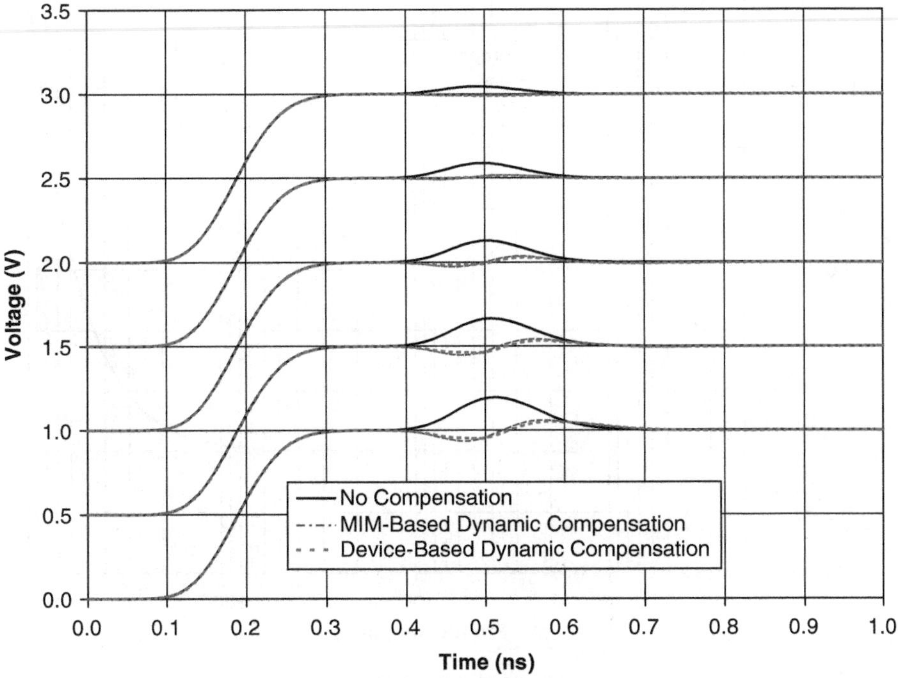

Fig. 15.7 Dynamic compensator TDR simulation results

Table 15.8 Reflection reduction due to dynamic compensator

$Length_{wb}$(mm)	$\Gamma_{No-Comp}(\%)$	$\Gamma_{MIM-Comp}(\%)$	$\Gamma_{Device-Comp}(\%)$	Setting
1	4.5	1.0	1.0	001
2	8.7	1.8	1.3	011
3	12.7	3.6	3.0	100
4	16.4	4.3	3.3	110
5	19.8	6.0	5.0	111

(3 GHz) input step. Again, the TDR waveforms are offset in the voltage axis for view-ability with the 1 mm curve on the top and the 5 mm curve on bottom. As in the case of the static compensator, we can observe that the dynamic compensator results in a dramatic reduction in reflections from the wire bond.

Table 15.8 reports the reduction in reflections when using the dynamic compensator(s), along with the binary control setting used for the compensation.

Figure 15.8 shows the input impedance of the wire bond structure versus frequency for the 3 mm wire bond using the dynamic compensator. Once again, adding a compensator can keep the lumped impedance of the wire bond closer to 50 Ω up to a much higher frequency. In this case, the 3 mm wire bond was kept to within 10 Ω of design up to 6.8 GHz when using the MIM-based compensator (compared to

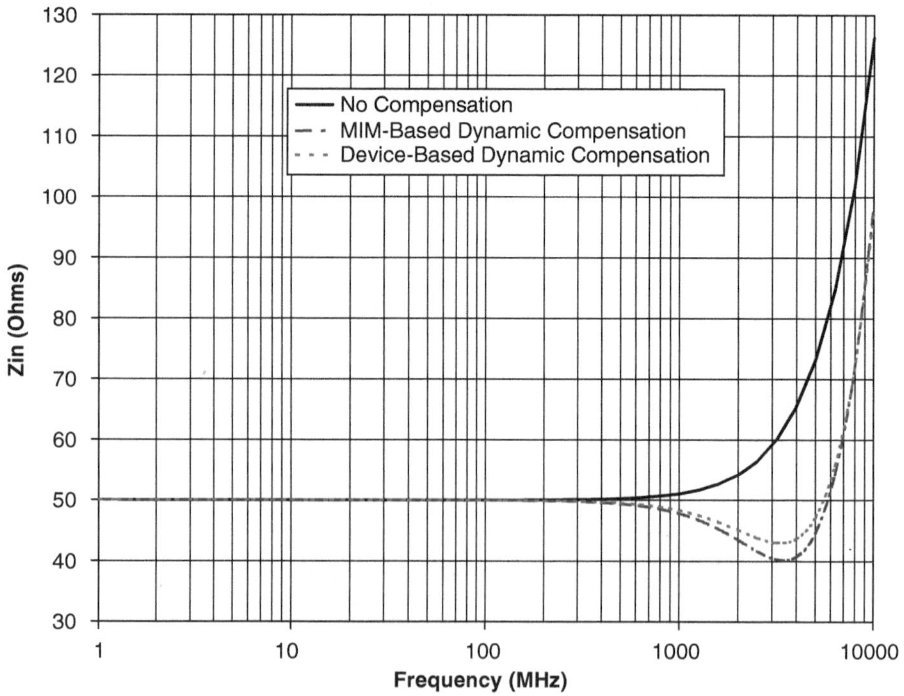

Fig. 15.8 Dynamic compensator input impedance simulation results

Table 15.9 Frequency at which dynamic compensator is +/−10 Ω from design

$Length_{wb}$(mm)	$f_{No-Comp}$(GHz)	$f_{MIM-Comp}$(GHz)	$f_{Device-Comp}$(GHz)	Setting
1	9.3	20	20	001
2	4.7	10.1	10	011
3	3.1	6.8	6.7	100
4	2.4	5.2	5.1	110
5	1.9	4.2	4.1	111

only 3.1 GHz when considering the uncompensated wire bond). The corresponding bandwidth of the Device-based compensator is 6.7 GHz.

Table 15.9 lists the frequencies at which the package impedance deviates +/−10 Ω from its target impedance of 50 Ω for all wire bond lengths, using the dynamic compensator.

Due to the nonlinearity of the Device-based capacitors and active pass gates, the variation of the dynamic compensator as a function of bias voltage was evaluated. For each dynamic compensator setting, the bias voltage was changed for $V_G = 0v$ to $V_G = 1.5v$, and the corresponding capacitance was recorded. Table 15.10 lists the effect of bias voltage on both the MIM-based and Device-based compensator circuits. This clearly shows the nonlinearity of the Device-based compensator which experiences as much as 33% capacitance variation when programmed to its maximum setting. This variation matches the expected variation of standard CMOS PolySilicon gate capacitors [59]. Note that the MIM capacitors also exhibit variability (3.8%), which occurs due to the bias dependence of the pass gate diffusion capacitances [2]. While both dynamic circuits exhibit a bias voltage dependence, both the compensators have sufficient range to cover wire bond lengths from 1 to 5 mm.

These results illustrate that the Device-based compensator outperformed the MIM-based compensator when implemented in a dynamic architecture. Also, both dynamic compensators outperformed their static counterpart. The dynamic compensator has the flexibility to be integrated in all future VLSI designs as part of the standard process to reduce the growing impact of package reflections.

Table 15.10 Dynamic compensator range and linearity

		MIM-Based Compensator			Device-Based Compensator		
Setting	$C_{(desired)}$ (fF)	$C_{(V_{bias}=0v)}$ (fF)	$C_{(V_{bias}=1.5v)}$ (fF)	$C_{average}$ (fF)	$C_{(V_{bias}=0v)}$ (fF)	$C_{(V_{bias}=1.5v)}$ (fF)	$C_{average}$ (fF)
001	200	252	262	257	222	281	251
010	325	373	382	378	318	414	366
011	450	499	540	519	423	587	505
100	575	588	596	592	485	651	568
101	700	713	754	734	592	816	704
110	825	828	895	862	688	968	828
111	950	948	1041	994	788	1180	984

15.2.4 Dynamic Compensator Calibration

To program the compensator to the optimal capacitance value, a calibration can be performed at package test.

The circuit used to perform the calibration is inspired by TDR ideas, but is simplified for applicability in a standard, low-cost VLSI tester setting. The calibration circuitry is simple and can be placed in the IC test equipment so that extra circuitry is not needed on the IC. In addition, pre-existing control logic and software interfaces in the tester can be used for calibration. Figure 15.9 shows the calibration circuitry for the compensator.

To measure the impedance discontinuity from the wire-bond/compensator, the IC tester transmits a voltage step into the IC package. The magnitude of the reflection from the discontinuity is measured in the tester using comparator circuits. The comparator circuits are used instead of a standard A/D converter (as in a true TDR system) to reduce complexity and cost. One comparator monitors for reflections that exceed a user-defined Upper Control Limit (UCL). A second comparator monitors for reflections that exceed a Lower Control Limit (LCL). The programmable voltages (UCL and LCL) are set by a Digital Control Monitor (DCM) in the IC tester.

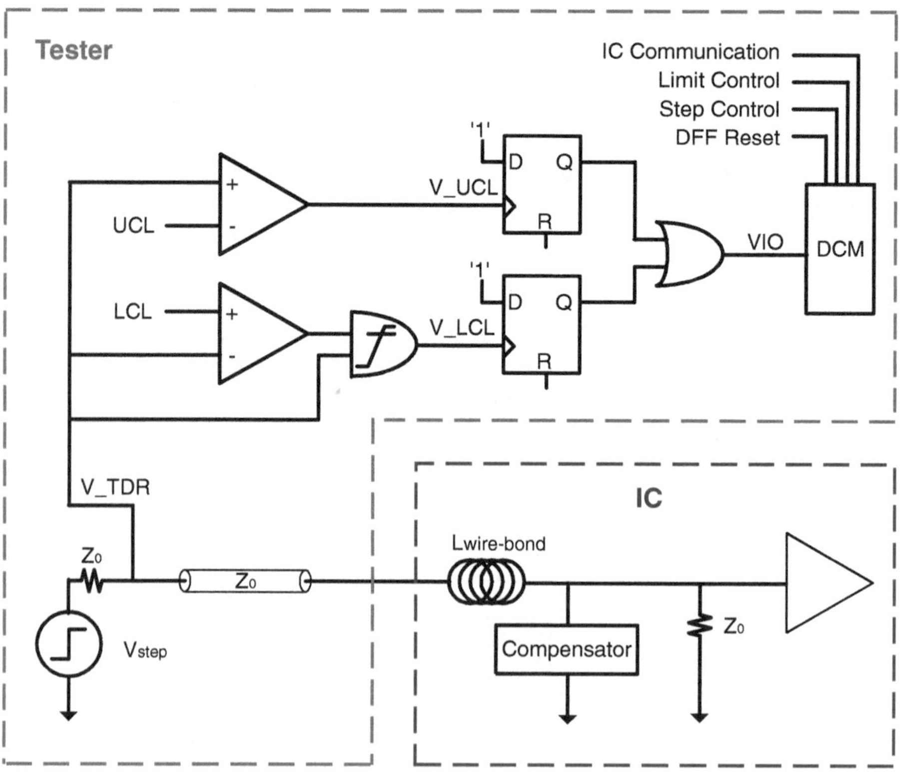

Fig. 15.9 Dynamic compensator calibration circuit

When a reflection from the wire-bond/compensator element is above/below the violation limits, the comparator(s) will output a glitch signal that indicates a limit violation (V_UCL / V_LCL). The limit violation glitch is fed into the clock input of a D-flip-flop whose data input is tied to a logic 1. The D-flip-flop therefore serves as a trigger element that will switch and remain high when it detects any glitch from the comparators. The D-flip-flop output will remain high until reset by the DCM. The output of the two D-flip-flops are fed into an OR-gate to combine the upper and lower violation signals into one input (VIO) that is monitored by the DCM.

When the DCM detects a violation, it will communicate with the IC under test to change the compensator settings. Once the compensator is adjusted, the D-flip-flops are reset, and another voltage step is launched into the IC package. This process is repeated until the magnitude of the reflections are within the LCL and UCL limits.

The lower control limit violation signal (V_LCL) is qualified using an AND-gate. The purpose of the AND-gate is to prevent glitches on V_LCL when the step voltage is below the LCL due to normal operating conditions such as the beginning of the rising edge of the step. The AND-gate is designed to have a high switch-point such that $V_{switchpoint-AND} > \mathrm{LCL}$. Once the step voltage exceeds the switch-point of the AND-gate, the glitches from the comparator are allowed to pass through to the D-flip-flop. Spurious glitches which may be observed when the step voltage is below the AND gate switch-point are thus filtered out, and the glitches at the output of the

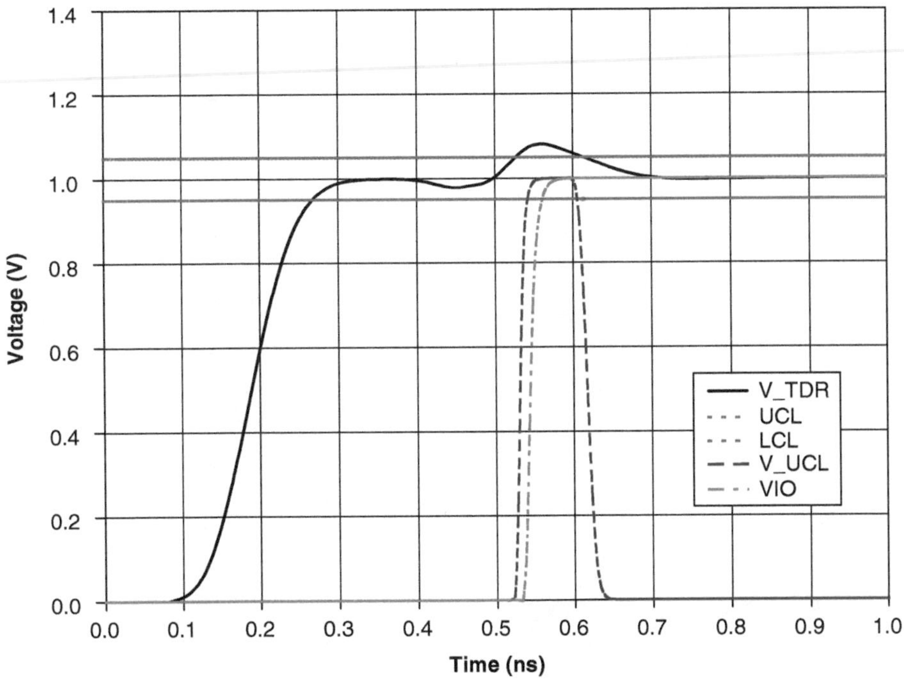

Fig. 15.10 Dynamic compensator calibration circuit operation

AND gate will therefore be solely due to reflections from the package (which violate the LCL).

In the case that the compensator cannot find a setting that meets the reflection control limits, it will relax the UCL and LCL voltages. Since this calibration is only performed on the signal pins of the IC package, the overhead associated with the process is small. The calibration can be sped up further by setting the UCL and LCL to predefined values that are based on the design of the IC and package. If the tester uses reasonable limits to begin with, the convergence of the algorithm will be sped up.

Figure 15.10 shows the signals of the calibration circuit during operation. In this case, the limits are set to indicate violations when reflections exceed 5%. In this figure, a UCL violation is indicated by a glitch on V_UCL. The glitch then triggers the D-flip-flop which in turn sends a static control signal (VIO) to the DCM.

Chapter 16
Future Trends and Applications

The analytical models and noise reduction techniques presented in Chaps. 11–15 were analyzed for use with past, present, and future IC packaging in order to predict and improve performance. The experimental results illustrated that the techniques were successful and made significant improvement in the performance of the packaging. While these techniques were demonstrated to have an immediate impact when applied to commonly used VLSI packages, the current trends in IC technology make these techniques even more invaluable. In addition, since all of the modeling and performance techniques were described using a common mathematical framework, the work in this monograph can be easily applied to a wide variety of electronic applications. This chapter presents some of the industry trends and other applications that may benefit from the modeling techniques described in this work.

16.1 The Move from ASICs to FPGAs

Application specific integrated circuits (ASICs) have enabled the dramatic increase in computational power that digital systems have enjoyed over the past 20 years; however, within the past 5 years, the cost associated with designing and fabricating an ASIC has increased considerably [40]. This cost stems from the increased complexity of the design process, along with the high cost of manufacturing that accompanies modern fabrication processes. As a consequence, there has been a marked reduction in the number of ASIC design starts per year compared to design starts 10 years ago. At the same time, Field Programmable Gate Arrays (FPGAs) have gained a foothold as one of the most-used building blocks in digital systems. The flexibility of an FPGA allows designers to reduce hardware design cycles times, while adding inherent feature upgradeability in the final product. Traditionally FPGAs have been used solely as a prototyping vehicle due to their higher cost and slower performance compared to ASICs; however, recently FPGAs have experienced a dramatic increase in performance due to the rapid improvement in IC processing technology. FPGA development has kept up with Moore's law, while at the same time providing inherent design flexibility. Figure 16.1 shows that the doubling of transistors in Intel microprocessors has followed the prediction of Moore's Law over the past 25 years

C. Duan et al., *On and Off-Chip Crosstalk Avoidance in VLSI Design*, DOI 10.1007/978-1-4419-0947-3_16,

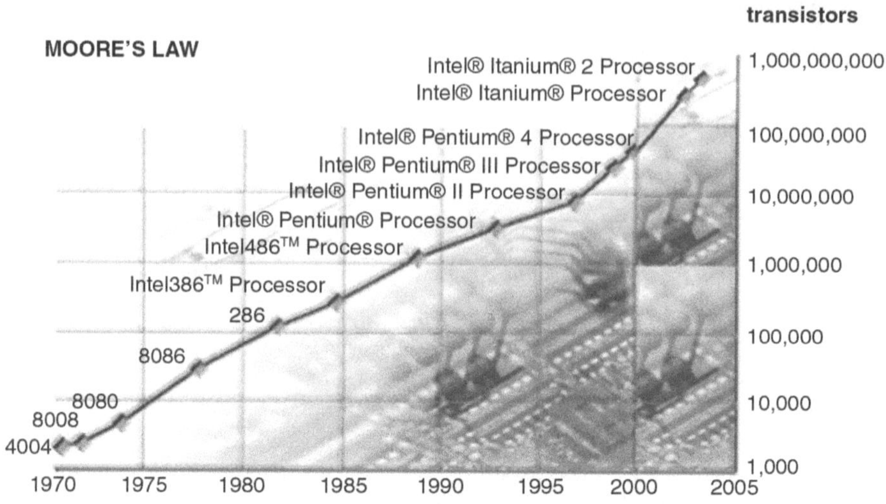

Fig. 16.1 Moore's law prediction chart

[55]. Figure 16.2 shows the recent increase in the amount of logic cells that can be implemented using an FPGA [101]. The rapid growth in FPGA technology illustrates how FPGAs are also tracking Moore's Law prediction.

Combining the increased performance of FPGAs with the dramatic increase in the price of ASICs, FPGAs are now the preferred implementation vehicle for digital systems not requiring extremely high performance. Within the past 5 years the number of digital design starts that use at least one FPGA is four times greater than the

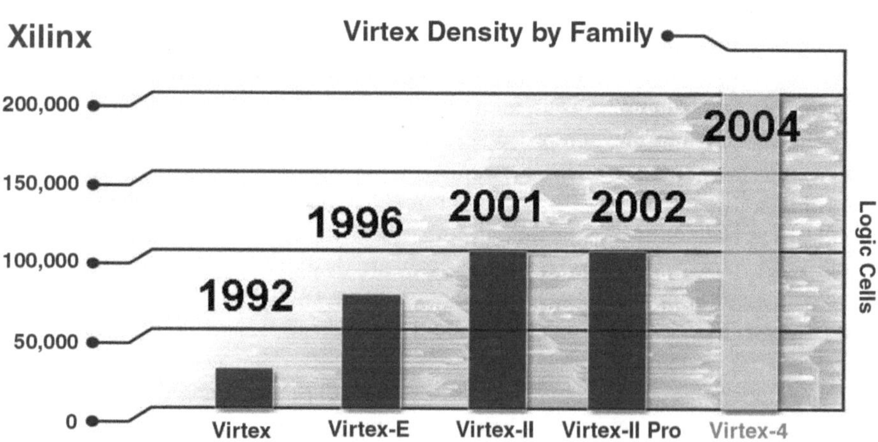

Fig. 16.2 Xilinx FPGA logic cell count evaluation

number of design starts requiring an ASIC [40]. This trend is important to this work due to the use model of FPGAs. An FPGA is designed for use in a wide variety of applications, which each present different electrical constraints. A single FPGA must operate over various supply voltages, system impedances, bus speeds, and I/O levels. In addition, a particular FPGA circuit design is typically implemented in many styles of packaging to offer a range of performance and cost to the users. This fact means that any technique that can improve system performance yet operate across many styles of packaging and bus configurations will be extremely valuable to the FPGA industry.

All of the techniques presented in this work can be applied to an FPGA. In fact, two of these techniques were prototyped using FPGAs as well. Since the analytical model and bus selection techniques presented in Chaps. 11 and 12 were constructed in terms of generic LC models of the package interconnect, they can be directly applied to any style of FPGA as long as the package parasitics are known. This enables system designers to quickly analyze and design off-chip busses as well as understand the impact of moving toward advanced packaging within an FPGA family.

The encoding techniques presented in Chaps. 13 and 14 can also be directly applied to FPGAs. Once a given bus configuration is implemented, performance can be further increased by implementing an encoding scheme to avoid the noise induced by bus data traversing the package parasitics. FPGAs lend themselves very well to the encoding methodology since logic can be added after design and fabrication. This allows encoders to be added *during* prototyping if simultaneous switching noise is found to be a problem. In addition, since the encoders can be added at anytime, they present the flexibility to address different noise sources depending on the type of packaging utilized. One FPGA in a system may have an inductive cross-talk noise problem while another FPGA may have an impedance discontinuity issue. Any source of noise that is of concern can be addressed by altering the encoder construction.

Finally, the compensator can also be used to directly improve performance in FPGAs. As mentioned earlier, an FPGA family is usually produced using the same design core, with various packaging, cost, and performance variants. This can be extremely challenging to VLSI designers who are faced with having to deal with multiple package parasitics for a single circuit. An FPGA circuit may exist in a wire bonded package which must be impedance matched to the system impedance. This is difficult to do since the same die may also be placed in a flip-chip package and must be impedance matched to the same system impedance. The dynamic compensator addresses this problem by having a programmable compensation capacitance that can be designed to cover a wide range of package parasitics. The dynamic compensator has the ability to match a wire bond package as well as a flip-chip package to a given system impedance by simply changing the control lines of the capacitive compensation.

All the techniques in this work favorably complement the trend of moving from ASICs to FPGAs. Since the manufacturing cost of ASICs is not predicted to lower in the foreseeable future, FPGAs will continue to dominate digital design starts. This indicates that the work in this monograph is applicable to a dominant industry trend that may continue for many decades.

16.2 IP Cores

Another recent trend in the VLSI industry is the move toward system design through use of *Intellectual Property* (IP) cores. An IP core is a circuit block that is designed and verified such that it can be dropped into a larger design. An IP core differs from standard ASIC designs in that it typically contains large, complex, system level circuitry. Blocks such as DDR memory controllers, PCI interfaces, and PowerPC microprocessors are examples of complex system-level IP cores. IP cores are also gaining acceptance as stand-alone products which are licensed to an end user who is designing an FPGA or ASIC. The move toward the IP core methodology comes from the ever-growing complexity of modern VLSI designs. It is becoming impractical for a single design team to design every block in an ASIC in a reasonable amount of time. As a consequence of time-to-market demands, design teams are turning to IP cores for the more standard portions of the ASIC.

The noise reduction techniques presented in this work lend themselves very elegantly to the IP core design methodology. Since each technique can be implemented using a general purpose block of circuitry (which is relatively independent of the IC design), noise reduction cores can easily be created and incorporated into larger VLSI designs. In fact, these cores can be parameterized, allowing them to be effective for more than one bus or package configuration. Bus encoding/decoding cores can be inserted between the core signals of the IC and package I/O pins. This core insertion would not have any impact on the core circuitry and serves only to improve the speed of the off-chip bus that traverses the package. Integrated circuits that are limited by the package noise can be sped up, so that the core logic on the die may be operated faster as a consequence.

Similarly, the compensator can also be implemented as an IP core. Since the compensator addresses the level 1 interconnect problem, the core circuitry of the die can be designed without the additional constraints that are driven by packaging speed limitations. In addition, since the compensator is shown to have sufficient range to cover a variety of package interconnects, it can be used for designs that go into multiple package technologies.

Figure 16.3 shows a possible scenario in which the techniques presented in this work can be integrated into a larger VLSI design as IP cores. In this example, two styles of off-chip busses are shown. The first is a *PCI* bus, which is implemented using a traditional wider, slower configuration. This style of bus experiences a large amount of supply bounce due to the large number of simultaneously switching signals. For this bus, an encoder is utilized to avoid inductive noise due to data sequences that traverse the package. The second off-chip bus uses a *HyperTransport* core. This bus is implemented using a faster, narrower configuration. In this configuration, the number of signals that simultaneously switch is considerably less than the PCI bus. For this example, the HyperTransport bus will not experience as much supply bounce but will instead be limited by capacitive bandwidth and impedance discontinuity problems. To address the bandwidth limitation, an encoder is utilized to avoid the worst-case bandwidth limiting sequences. To address the impedance discontinuities, a compensator circuit is used to impedance match the package interconnect with the

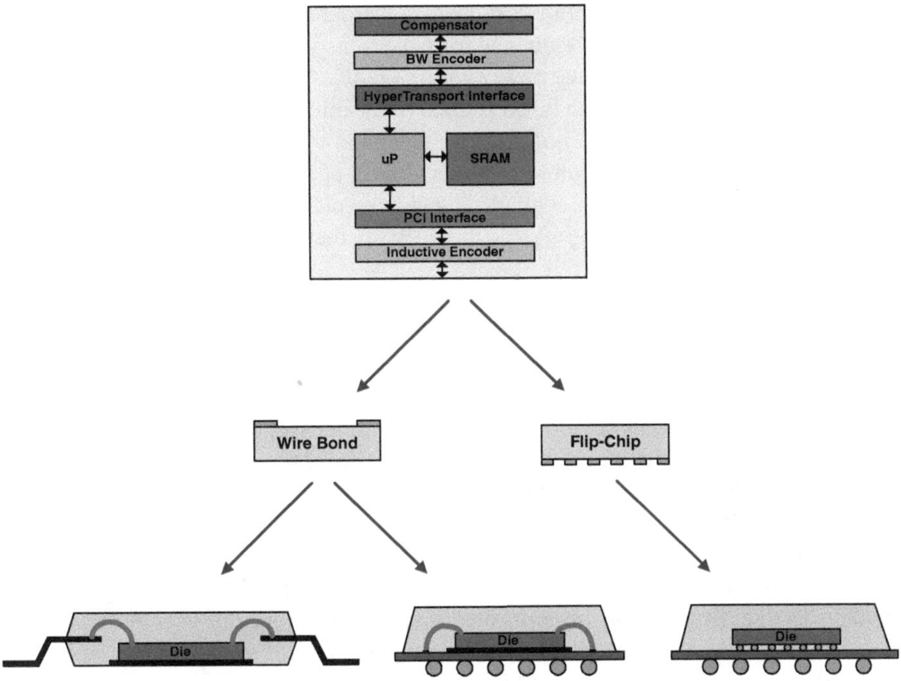

Fig. 16.3 IP core design methodology incorporating encoder and compensator

system PCB. Since the noise reduction cores can be implemented in a parameterized manner, variations in the code construction allow the core to be usable in a variety of packaging and bus configurations.

16.3 Power Minimization

Another major challenge for VLSI designers in this and the next decade is how to reduce the extreme amount of power that is being consumed as more transistors are integrated on-chip. Advances in IC processing technology enables more transistors to be integrated on a die than can be cooled using modern packaging techniques. Additionally, the ever-diminishing transistor feature sizes enable faster switching times for the core logic. Equation 16.1 expresses the dependence of power consumption on the capacitance (C_{load}) being switched, supply voltage (V_{DD}), switching frequency (f_{switch}), and the activity factor (α_{switch}).

$$P_{VLSI} = (C_{load}) \cdot (f_{switch}) \cdot (V_{DD})^2 \cdot (\alpha_{switch}) \tag{16.1}$$

The switching activity factor, α_{switch}, represents the fraction of circuit nodes that switch during any clock cycle. This parameter is typically between 0.2 and 0.35 for modern VLSI designs [59].

The encoders presented in this work were constructed in a general manner so as to eliminate any arbitrary vector sequence from the transmitted data. This methodology can be easily extended to reduce power within the IC. As mentioned earlier, as much as 25–50% of the IC power can be consumed in the output drivers [11, 40]. Since the encoders described in this monograph are designed to eliminate vector transitions that result in noise of a magnitude greater than a specified value, these encoders can also be used to avoid data sequences which result in large power consumption. This has the effect of reducing α_{switch} in Eq. 16.1, thereby directly reducing power consumption.

A further extension of the encoder can be used to avoid vector sequences in on-chip busses which results in a power consumption above a specified limit. Long on-chip busses suffer not only an increased power consumption due to their typically high switching activity, but also from the fact that the bus capacitance (C_{load}) is proportional to bus length. The problem of long on-chip busses consuming considerable power has become more important as more system level blocks are integrated within one die. To incorporate complex system blocks, the die size increases, creating a need for long block-to-block busses. These busses are inherently long therefore consume considerable power, according to Eq. 16.1. Implementing encoders that reduce the worst-case power consumption sequences in long on-chip busses will have a direct impact on total power consumption of the IC. In addition, the encoder construction methodology needs minimal modification, with addition constraint equations required to express the user-defined power limits.

Since power is a critical issue in current and future designs, it is often conjectured that power will eventually limit the continued success of Moore's Law. In such a scenario, power reduction techniques like the one described above may play an important part in extending the success of Moore's Law

16.4 Connectors and Backplanes

Nearly all digital systems contain connectors which transmit signals and power between various printed circuit boards. Connectors provide both the mechanical and electrical connection between PCBs. Connector construction is similar in nature to that of IC packaging in that the mechanical robustness of the interconnect must be addressed before considering the electrical behavior. Once a stable mechanical structure is designed that can be economically manufactured, then the electrical performance can be evaluated. This results in backplane interconnect that is not optimized for electrical performance. This is the exact situation that exists in IC packaging. Connector leads contain electrical parasitics which degrade the performance of digital signals. Excess inductance and capacitance in the connector leads to voltage drop, capacitive bandwidth limitation, signal-to-signal coupling, and impedance discontinuities.

Since all the techniques presented in this work are constructed to alleviate the detrimental effects of the inductance and capacitance of the interconnect, this naturally makes them applicable to any noise-prone electrical interconnect in a digital system. The analytical modeling, encoding scheme, and compensation methodology presented in this monograph can be applied to any connector without modification. Instead of using the inductance and capacitance parasitics from the IC package, the inductance and capacitance of the connector would be used. From these values, the noise associated with the electrical parasitics of the connector can be directly computed. In addition, the encoding and compensation techniques can be used to bound the maximum amount of connector noise.

The application of the noise reduction techniques to connectors plays well into the *backplane* architecture used throughout industry. The backplane architecture is used widely throughout the computer industry as a way to scale systems. Smaller circuit blocks are implemented which can be plugged into a main PCB to add computational power as needed. Examples of circuit blocks that utilize this architecture are microprocessors, memory, and non-volatile storage elements. The backplane methodology allows the smaller circuit blocks (which are utilized in the larger system) to be designed and verified independent of the larger system. This results in a partitioning of the system functionality and complexity, resulting in reduced design cycle times and more robust system operation. In addition, the smaller circuit blocks can be produced by multiple vendors, which drives down the cost of the design. The advantages of the backplane architecture have resulted in a widespread adoption of this methodology. This architecture has in turn driven the performance of some of the most popular off-chip bus standards. Busses such as PCI Express, Infiniband, and Gigabit Ethernet were designed specifically to improve throughput of backplanes. These busses are predicted to achieve per-pin datarates of 10Gb/s within the next five years [11, 77]. Since the techniques presented in this work apply directly to the connectors used in the backplane methodology, their importance could potentially play a crucial part in the continued increase in performance of backplane-based system.

16.5 Internet Fabric

One of the most important developments in the past 30 years has been the internet. This technology has revolutionized communication and sparked one of the longest growth periods for the global economy in recent history. One of the underlying technologies that has enabled the success of the internet is the *internet protocol*. The internet protocol allows communication data to be broken into smaller pieces and transmitted as a series of packets, each of which contains destination address information. The protocol allows data to be sent through multiple infrastructure paths and be reassembled by the receiving computer. Packets may arrive out of order on account of varying delays along different electrical paths. Various alternative infrastructure paths are selected according to the usage on any given path.

Many encoding protocols have been developed over the past decade to address the problem of congestion within the internet infrastructure (or fabric) [12, 48, 92]. These protocols attempt to avoid network congestion by monitoring the sequence of data packets that are being transmitted. In applications such as audio and video, encoding algorithms monitor the data stream for adjacent packets that contain identical information. In an A/V application, sequential data patterns that contain the same information may be eliminated without noticeable distortion in the final reassembled data. These encoding algorithms allow streaming audio and video to be transmitted reliably and efficiently despite congested network fabrics, to be eventually received as a reasonable representation of the original data.

The encoders presented in this work are constructed by creating a directed graph containing transitions that have been deemed *legal* by the constraint equation evaluations. In the package noise constraints within this work, transitions are eliminated from the complete set of encoder transitions if they violate any of the user-defined noise limits for the system; however, the constraint equations can be easily modified to detect sequences of data patterns that can be eliminated in an A/V application when faced with network congestion. This modification would consist of a new series of constraint equations that are written specifically for the A/V application. With the increasing use of streaming audio and video over the internet, the encoding techniques presented in this work could potentially provide significant improvement in the throughput of A/V data over congested network paths.

This chapter presented potential applications of the techniques in this work to current and future industry trends. In each case, the general manner in which the noise reduction techniques are formulated enables their use in a variety of applications. Any system which experiences unwanted noise due to the electrical parasitics of the interconnect or simply desires the elimination of specific transitions can benefit from the techniques presented in this monograph.

Chapter 17
Summary of Off-Chip Crosstalk Avoidance

This monograph has presented a comprehensive look at the noise problems within IC packaging. Today's integrated circuits are experiencing a dramatic increase in performance due to significant advances in the IC design and fabrication processes. IC technology has followed Moore's Laws for the past 30 years and is expected to continue at this rate sometime into the next decade. At the same time, IC packaging technology has evolved at a much slower pace. This mismatch in performance between the IC and the package is now the leading limitation to system performance. Inter-chip busses that transfer data between ICs within the system need to be slowed so as to avoid unwanted noise from the package. Off-chip communication is now the largest bottleneck in modern digital systems design.

Unwanted noise occurs in the package due to the parasitic inductance and capacitance within the package interconnect. The parasitics of the package can cause supply bounce, signal-to-signal coupling, capacitive bandwidth limiting, and impedance discontinuities. Since the IC packaging was originally developed for robust mechanical performance and manufacturability, the interconnect was not optimized for electrical performance. In addition, the complex manufacturing processes used to create the package make altering the interconnect to improve electrical performance very difficult and expensive. The move toward advanced packaging can reduce the electrical parasitics in the package but is often too expensive for the majority of VLSI designs. As such, any technique that can aid VLSI designers in the prediction of performance and reduction of noise within the package is of great value.

Chapter 11 presented an analytical model to predict the performance of an IC package. This was accomplished by finding the fastest rate of change in current or voltage that could be tolerated without violating any of the user-defined noise limits. The amount of noise in the package is dependant on how large the parasitic inductance and capacitance are within the package interconnect. The electrical rate of change of current/voltage was translated a risetime figure and, in turn, into per-pin datarate and bus throughput. For each of the sources of noise, equations were derived that considered the package parasitics and the user-defined noise limits. To verify the accuracy of the models, SPICE simulations were performed on a test circuit. It was found that the analytical model matched simulated results within 10% for bus segments up to 16-bits. It was also found that for the three packages studied

C. Duan et al., *On and Off-Chip Crosstalk Avoidance in VLSI Design*,
DOI 10.1007/978-1-4419-0947-3_17, © Springer Science+Business Media LLC 2010

in this work, the inductive supply bounce and inductive signal coupling were the dominant noise sources. The analytical models derived in this chapter provide a method for VLSI designers to quickly analyze different packaging technologies and bus configurations in order to meet their throughput requirements.

Chapter 12 presented a technique to select the most cost-effective bus configuration. Using the analytical models from Chap. 11 and including the per-pin cost of the bus configuration, an algorithm was developed that could easily determine the most cost-effective package selection and configuration. In addition, the metric of Bandwidth-per-Cost was introduced, which quantified the cost-effectiveness of a given bus. Using this metric, the total cost of different bus and package configurations (all of which meet the desired throughput) can be compared efficiently. The algorithm gives VLSI designers a powerful deterministic tool to find the most cost-effective off-chip bus design.

Chapter 13 presented a bus expansion encoding technique that was able to reduce package noise by encoding the data prior to it traversing the package interconnect. The encoder algorithm consisted of a series of constraint equations which were written to model noise violations for arbitrary bus transitions. These equations were evaluated to indicate if a given transition would result in a noise limit violation when traversing the package interconnect. Using the remaining legal transitions, a directed graph was created that was used to construct the encoder. Each vertex within the graph was evaluated in turn to find if its remaining outgoing edges were able to encode the transitions of an m-bit bus. The output of the encoder is a mapping of transitions in the original m-bit bus to transitions in an expanded n-bit bus, such that the transitions in the expanded bus enable data transmission without noise violations.

Since the encoded data introduces less package noise, it can be transmitted at a higher per-pin datarate. This speed increase results in throughput improvement even after accounting for the encoder overhead. Experimental results performed on the encoder illustrated that for a fixed $\frac{di}{dt}$ of $33\frac{MA}{s}$, package noise was reduced as much as 89% for an aggressively encoded bus. It was also shown that for a 3-bit bus which was encoded and evaluated using a varying $\frac{di}{dt}$, the bus throughput was increased up to 46% compared to the original unencoded data. The bus expansion encoding technique presented in this work provides VLSI designers a complete methodology to reduce the noise within the IC package and increase off-chip bus throughput.

Chapter 14 presented a bus stuttering encoder which was also successful in reducing the amount of package noise. This technique was similar to the bus expansion encoder in terms of the creation and evaluation of the constraints and legal transitions of the bus; however, a stuttering algorithm was invoked on the directed graph of legal transitions, which inserted intermediate states between pairs of vertices which resulted in noise limit violations. The stutter states were inserted such that each vertex of the directed graph could transition to any other vertex using only the remaining legal vectors. This encoder technique had the advantage that no additional package pins were needed to encode the m-bit bus. This technique did require more area on the IC than the bus expansion encoder, but was deemed to be incremental to existing protocol-based busses that already contain state machines in their output circuitry. Experimental results illustrated that throughput could be increased as much as 225% for a 6-bit, aggressively encoded bus. The bus stuttering encoder presented in this work provides VLSI designers with a technique to improve throughput in their off-chip busses without adding pins to the package.

Chapter 15 presented a compensation technique that was able to match the impedance of the package interconnect to that of the system PCB. This was accomplished by adding additional capacitance near the wire bonds or flip-chips within the package. This has the effect of lowering the typically high interconnect impedance to a level that matches the system PCB impedance. By matching the impedance of the package to the system PCB, reflections are avoided and performance can be increased. Static and dynamic compensation techniques were presented. In the static approach, predefined capacitance was placed on-chip and on the package to surround the inductive level 1 interconnect. The static compensator was able to reduce reflections in a 5 mm wire bond from 20% to 5% for a risetime of 117 ps. In the dynamic approach, a programmable capacitance was placed on-chip that could be altered after IC fabrication. The dynamic technique allowed the same circuit to compensate for various interconnect inductances which may result from design or process variations in the interconnect. It was demonstrated that the dynamic technique was able to reduce reflections of a 5 mm wire bond from 20% to 6% for a risetime of 117 ps. The compensators presented in this work offer VLSI designers a simple technique to impedance match the package interconnect to the system PCB without noticeable area utilization within the package or on the die.

Chapter 16 gave an overview of other applications to which the techniques presented in this work are well suited. The move from ASICs to FPGAs was described, along with a discussion of how the noise reduction techniques reported in this work apply directly to the FPGA design methodology. In addition, the utility of implementing the techniques as IP cores was discussed. The application of the encoders to the problem of power reduction within the IC was also presented. Finally, the possible use of these techniques to connectors, backplanes, and the internet fabric was pointed out. While only a subset of applications were presented in Chap. 16, it is believed that the techniques in this monograph can be applied to a wide variety of electronic applications due to their flexible formulation.

With the continued advancement in IC technology, the performance limitation of the package will continue to dominate system performance. While advanced packaging can aid in reducing the noise associated with the parasitics in the package, it is often too expensive for deployment in the majority of VLSI designs. Part II or this monograph presented noise analysis and noise reduction techniques that can be directly applied to current and future packaging technologies. With the use of the techniques presented in this monograph, the VLSI community can continue to experience the dramatic increases in the computational power of digital systems.

References

[1] Berkeley predictive technology modeling homepage. http://www-device.eecs.berkeley.edu/bptm/.

[2] BSIM4 offcial release site. http://www-device.eecs.berkeley.edu/bsim3/bsim4.html.

[3] Fibonacci number (From Wikipedia). http://en.wikipedia.org/wiki/Fibonacci_number.

[4] Moore's law. http://en.wikipedia.org/wiki/Mooreś_law.

[5] Numeral system. http://en.wikipedia.org/wiki/Numeral_system.

[6] Physical Design Modelling and Verification Project (SPACE Project). http://cas.et.tudelft.nl/research/space/html.

[7] TSMC 65 nm process. http://www.tsmc.com/english/b_technology/b01_platform/b010101_65nm.htm.

[8] TSMC 90 nm process. http://www.tsmc.com/english/b_technology/b01_platform/b010101_90nm.htm.

[9] Xilinx VIRTEX4 family datasheet. http://www.xilinx.com/products/silicon_solutions/fpgas/virtex/virtex4/index.htm".

[10] 1958: The invention of the integrated circuit. Technical report, www.pcb.org, 2002.

[11] The international technology roadmap for semiconductors (itrs). Technical report, public.itrs.net, 2003.

[12] A.S. Abraham, J. Wang, and J.C.L. Liu. Bandwidth-aware video encoding with adaptive image scaling. In *IEEE International Conference on Multimedia and Expo (ICME)*, pages 157–160, June 2004.

[13] J.H. Ahm, K.T. Lee, M.K. Jung, Y.J. Lee, B.J. Oh, S.H. Liu, Y.H. Kim, Y.W. Kim, and K.P. Suh. Integration of MIM capacitors with low-k/Cu process for 90 nm analog circuit applications. In *Proceedings of the IEEE International Interconnect Technology Conference*, pages 183–185, June 2003.

[14] N. Arora, K. Raol, R. Shcumann, and L. Richardson. Modeling and extraction of interconnect capacitance for multilayer VLSI circuits. In *IEEE Transactions on Computer-Aided Design of Integrated Circuits and Systems*, 15(1):58–67, 1996.

[15] Avant! Raphael: Interconnect Analysis Program User's Guide. www.avant!.com/raphael, 2000.

[16] D. Balaraman, J. Choi, V. Patel, P.M. Raj, I.R. Abothu, S. Bhattacharya, L. Wan, M. Swaminathan, and R. Tunimala. Simultaneous switching noise suppression using hydrothermal barium titanate thin film capacitors. In *Proceedings of Electronic Components and Technology (ECTC)*, pages 282–288, June 2004.

[17] A. Bastani and C.A. Zukowski. A low-leakage high-speed monotonic static CMOS 64b adder in a dual gate oxide 65-nm CMOS technology. In *Proceedings of International Symposium on Quality Electronic Devices*, pages 312–317, 2006.

[18] G.E. Beers and L.K. John. A novel memory bus driver/receiver architecture for higher throughput. In *Proceedings of the Eleventh International Conference on VLSI Design: VLSI for Signal Processing*, pages 259–264. IEEE, 1998.

[19] B. Bollobas. *Graph Theory*. Springer-Verlag, New York, 1979.

[20] K.S. Brace, R.L. Rudell, and R.E. Bryant. Efficient implementation of a BDD package. In *Design Automation Conference (DAC)*, pages 40–45, June 1990.

[21] R.N. Bracewell. *The Fourier Transform and its Applications*. McGraw-Hill, 2000.

[22] R.E. Bryant. Graph based algorithms for Boolean function representation. In *IEEE Transactions on Computers*, volume C-35, pages 677–690, August 1990.

[23] B. Casper. An accurate and efficient analysis method for multi Gb/s chip-to-chip signaling schemes. In *Symposium of VLSI Circuits Digest of Technical Papers*, pages 54–57, June 2002.

[24] B. Casper, A. Martin, J.E. Jaussi, J. Kennedy, and R. Mooney. An 8-Gb/s simultaneous bidirectional link with on-die waveform capture. In *IEEE Journal of Solid-State Circuits*, volume 38, pages 2111–2120, December 2003.

[25] J. Chem, J. Huang, L. Aldredge, P. Li, and P. Huang. Multilevel metal capacitance models for CAD symbol design synthesis systems. In *IEEE Electron Device Letter*, 13:32–34, 1992.

[26] C.L. Chen and B.W. Curran. Switching codes for delta-I noise reduction. In *IEEE Transactions of the 43rd IEEE Midwest Symposium on Circuits and Systems*, volume 45, pages 1017–1021, September 1996.

[27] K.Y. Chou and M.J. Chen. Active circuits under wire bonding I/O pads in 0.13 μm eight-level Cu metal, FSG low-k inter-metal dielectric CMOS technology. In *IEEE Electron Device Letters*, pages 466–468, October 2001.

[28] U. Choudhury and A. Sangivanni-Vincetelli. Automatic generation of analytical models for interconnect capacitance. In *IEEE Transactions on Computer-Aided Design*, 14:470–480, 1995.

[29] M.D. Ciletti. *Modeling, Synthesis, and Rapid Prototyping with the Verilog HDL*. Prentice Hall, 1999.

[30] HyperTransport Consortium. Technical report, www.hypertransport.org, 2005.

[31] W.H. Dally and J. Poulton. Transmitter Equalization for 4-Gbps Signaling. In *IEEE Micro*, volume 17, pages 48–56, February 1997.

[32] W.H. Dally and J. Poulton. *Digital Systems Engineering*. Cambridge University Press, Cambridge, U.K., 1998.

[33] S. Das. *Design Automation Techniques for Datapath Circuit*. PhD thesis, University of Colorado, Boulder, Colorado, 2007.

[34] C. Duan, K. Gulati, and S.P. Khatri. Memory-based cross-talk canceling CODECs for on-chip buses. In *Proceedings of International Symposium on Circuits and Systems*, pages 1119–1122. IEEE, May 2006.

[35] C. Duan and S.P. Khatri. Exploiting crosstalk to speed up on-chip buses. In *Proceedings of Design Automation, and Test in Europe (DATE) Conference*, 2004.

[36] C. Duan, A. Tirumala, and S.P. Khatri. Analysis and avoidance of cross-talk in on-chip bus. In *Proceedings of the The Ninth Symposium on High Performance Interconnects*, pages 133–138, 2001.

[37] SEMICON Far East. Technical report, www.semiconfarest.com, 2005.

[38] I.M. Elfadel. On-chip bus interleaving revisited. In *Proceedings of IEEE 14th meeting on electrical performance of electronic packaging*, 2005.

[39] M. Ghoneima and Y. Ismail. Optimum positioning of interleaved repeaters in bidirectional buses. *Transactions on Computer-Aided Design*, 24(3):461–469, 2005.

[40] Agilent Technologies, Inc. Packaging Group. Personal communication, 2004. Ft. Collins, CO.

[41] R. Gupta. Deep-submicron challenges. In *IEEE Design and Test of Computers*, 2002.

[42] L. He and K.M. Lepak. Simultaneous shield insertion and net ordering for capacitive and inductive coupling minimization. In *Proceedings of International Symposium on Physical Design*, pages 55–60.

[43] L. He and M. Xu. Characteristics and modeling for onchip inductive coupling. Technical Report ECE-00-1, Univeristy of Wisconsin at Madison, Madison, WI, 2003.

[44] K. Hirose and H. Yasuura. A bus delay reduction technique considering crosstalk. In *Proceedings of the conference on Design, automation and test in Europe*, pages 441–445, Paris, France.

[45] R. Ho, K. Mai, and M. Horowitz. Efficient on-chip global interconnects. In *Proceedings of Symposium on VLSI Circuits*, pages 271–274, 2003.

[46] J.M. Hobbs, H. Windlass, V. Sundaram, S. Chun, G.E. White, M. Swaminathan, and R.R. Tummala. Simultaneous switching noise suppression for high speed systems using embedded decoupling. In *Proceedings of the 51st Electronic Components and Technology Conference*, pages 339–343, June 2001.

[47] M. Horowitz, C. Yang, and S. Sidiropoulos. High-speed electrical signaling: Overview and limitations. In *IEEE Micro.*, volume 18, pages 12–24, January 1998.

[48] A.A.M. Ibrahim. Statistical rate control for efficient admission control of MPEG-2 VBR video sources. In *IEEE Proceedings of the ATM Workshop*, pages 26–29, May 1998.

[49] Actel Inc. Simultaneous switching noise and signal integrity. Technical report, www.actel.com, 2004.

[50] Agilent Technologies Inc. Advanced design systems user's manual. http://www.agilent.com/ads, 2000.

[51] ASAT Inc. Peak Performance Array Packages. Technical report, www.asat.com, 2004.

[52] ASAT Inc. Peak Performance Enhanced Lead Packages. Technical report, www.asat.com, 2004.

[53] ASAT Inc. Peak Performance Flip Chip Packages. Technical report, www.asat.com, 2004.

[54] Intel Inc. Intel itanium 2 processor. Technical report, www.intel.com, 2003.

[55] Intel. Prediction to reality. Technical report, www.intel.com, 2005.

[56] Y. Iraqi, R. Boutaba, and R. Dssouli. Statistical properties of MPEG video traffic and their impact on bandwidth allocation in wireless ATM networks. *IEEE Wireless Communications and Networking Conference (WCNC)*, pages 998–1002, September 1999.

[57] H. Johnson and M. Graham. *High-Speed Digital Design*. Prentice Hall PTR, 2003.

[58] H. Johnson and M. Graham. *High-Speed Signal Propagation*. Prentice Hall PTR, 2003.

[59] S. Kang and Y. Lebledici. *CMOS Digital Integrated Circuits, 2nd edition*. McGraw-Hill Companies, 1999.

[60] H. Kaul, D. Sylvester, and D. Blauuw. Active shielding of RLC global interconnects. In *International workshop on Timing issues in the specification and synthesis of digital systems*.

[61] S.P. Khatri. *Cross-talk Noise Immune VLSI Design using Regular Layout Fabrics*. PhD thesis, University of California at Berkeley, Berkeley, California, 1999.

[62] S.P. Khatri, A. Mehrotra, R.K. Brayton, Ralf H.J.M. Otten, and A.L. Sangiovanni-Vincentelli. A novel vlsi layout fabric for deep sub-micron applications. In *Proceedings of Design Automation Conference*, pages 491–496. IEEE, 1999.

[63] H. Kim, Y. Jeong, J. Park, S. Lee, and J. Hong. Significant reduction of power/ground inductive impedance and simultaneous switching noise by using embedded film capacitor. *Transactions on Electrical Performance of Electronic Packaging*, pages 129–132, October 2003.

[64] H. Kim, B.K. Sun, and J. Kim. Suppression of GHz range power/ground inductive impedance and simultaneous switching noise using embedded film capacitors in multilayer packages and PCBs. In *IEEE Microwave and Wireless Components Letters*, volume 14, pages 71–73, February 2004.

[65] K. Kim, K. Baek, N. Shanbhag, C. Liu, and S. Kang. Coupling-driven signal encoding scheme for low-power interface design. In *Proceedings of International Conference on Computer Aided Design*, pages 318–321. IEEE/ACM, 2000.

[66] C. Kinnaird. Standards are key to optimizing high-speed data bus communications. *Planet Analog (planetanalog.com)*, October 2002.

[67] L. Li, N. Vijaykrishnan, M. Kandemir, and M.J. Irwin. A crosstalk aware interconnect with variable cycle transmission. In *Proceedings of Design automation and test in Europe*, pages 102–107, 2004.

[68] J. Lin. Challenges for socdesign in very deep submicron technologies. International Conference on Hardware/Software Codesign and System Synthesis, 2003.

[69] M. Lopez, J.L. Prince, and A.C. Cangellaris. Influence of a floating plane on effective ground plane inductance in multilayer and coplanar packages. In *IEEE Transactions on Advanced Packaging*, volume 22, pages 182–188, May 1999.

[70] A.C.W. Lu, W. Fan, L. Wai, C.K. Wang, and H.G. Low. Design optimization of wire bonding for advanced packaging. In *Proceedings of Electronic Components and Technology Conference*, pages 1364–1372. ACM/IEEE, May 2003.

[71] J. Madsen and S. Long. A high-speed interconnect network using ternary logic. In *Proceedings of 25th International Symposium on Multi-Value Logic*. IEEE, January 1995.

[72] C. Mattei and A.P. Agrawal. Electrical characterization of BGA packages. In *Proceedings of the 47th Electronic Components and Technology Conference*, pages 1087–1093, May 1997.

[73] E. Mejia-Motta, F. Sandoval-Ibarra, and J. Santana. Design of CMOS buffers using the settling time of the ground bounce voltage as a key parameter. In *Proceedings of the 43rd IEEE Midwest Symposium on Circuits and Systems*, volume 2, pages 718–721, August 2000.

[74] M. Miura, N. Hirano, Y. Hiruta, and T. Sudo. Electrical characterization and modeling of simultaneous switching noise for leadframe packages. In *Proceedings of the 45th Conference on Electronic Components and Technology*, pages 857–864, May 1995.

[75] M. Mutyam. Preventing crosstalk delay using Fibonacci representation. In *International Conference on VLSI Design*, pages 685–688, January 2004.

[76] L. Nagel. Spice: A computer program to simulate computer circuits. In *University of California, Berkeley UCB/ERL Memo M520*, May 1995.

[77] PCI-SIG Trade Organization. Technical report, www.pcisig.com/, 2004.

[78] S. Palnitkar. *Verilog HDL - A Guide to Digital Design and Synthesis*. SunSoft Press, 1996.

[79] M.D. Powell and T.N. Vijaykumar. Pipeline damping: a microarchitectural technique to reduce inductive noise in supply voltage. In *Proceedings of the 30th International Symposium on Computer Architecture*, pages 72–83, June 2003.

[80] J. Prasad, M. Anser, and M. Thomason. Electrical characterization of dielectrics (oxide, nitride, oxy-nitride) for use in MIM capacitors for mixed signal applications. In *International Semiconductor Device Research Symposium*, pages 326–327, December 2003.

[81] C. Raghunandan, K.S. Sainarayanan, and M.B. Srinivas. Process Variation Aware Bus-Coding Scheme for Delay Minimization in VLSI Interconnects. In *Proceedings of International Symposium on Quality Electronic Devices*, pages 43–47, 2008.

[82] Rapid IO Trade Association. Technical report, www.rapidio.org/, 2004.

[83] O. Rose. Statistical properties of MPEG video traffic and their impact on traffic modeling in ATM systems. In *Proceedings from the 20th Conference on Local Computer Networks*, pages 397–406, October 1995.

[84] T. Sakurai. Design challenges for 0.1 um and beyond. In *Proceedings of the Asia South Pacific Design Automation Conference (ASP-DAC)*, pages 553–558, 2000.

[85] S. Haghani, and Bernard. A low power design of gray and T0 codecs for the address bus encoding for system level power optimization. In www.studentimaster.usilu.net/saraswap/prabhat/projects/Low_power_presentation.pdf.

[86] N. Satyanarayana, M. Mutyam, and A.V. Babu. Exploiting On-Chip Data Behavior for Delay Minimization. In *Proceedings of International Workshop on System-Level Interconnect Prediction*, Austin, Texas, USA, 2007. IEEE/ACM.

[87] D. Schinkel, E. Mensink, E.A.M. Klumperink, E. van Tuijl, and B. Nauta. A 3-gb/s/ch transceiver for 10-mm uninterrupted RC-limited global on-chip interconnects. *IEEE* Journal of Solid-State Circuits, 41(1):297–306, 2006.

[88] P.P. Sotiriadis and A. Chandrakasan. Low power bus coding techniques considering inter-wire capacitance. In *Proceedings of Custom Integrated Circuits Conference*, pages 507–510. IEEE, 2000.

[89] S.R. Sridhara, A. Ahmed, and N.R. Shanbhag. Area and energy-efficient crosstalk avoidance codes for on-chip buses. In *Proceedings of the International Conference on Computer Design*, pages 12–17. IEEE, 2004.

[90] S.R. Sridhara and N.R. Shanbhag. Coding for system-on-chip networks: a unified framework. In *Proceedings of Design Automation Conference*, pages 103–106. IEEE/ACM, 2004.

[91] M. Stan and W. Burleson. Bus-invert coding for low-power i/o. In *IEEE Transactions on Very Large Scale Integration (VLSI) Systems*, pages 49–58, 1995.

[92] A. Sugiura, M. Kamata, and A.T. Hayashi. MPEG video encoding based on assigning a high information priority to the focused region. *Asia-Pacific Conference on Circuits and Systems (APCCAS)*, pages 545–548, October 2002.

[93] D. Sylvester and K. Keutzer. Getting to the bottom of deep submicron. In *Proceedings of International Conference on Computer-Aided Design*, pages 203–211. IEEE/ACM, 1998.

[94] G.M. Tarolli and W.J. Herman. Hiearchical circuit extraction with detailed parasitic capacitance. In *Proceedings of Design Automation Conference*, 1983.

[95] A. Tucker. *Applied Combinatorics, 3rd edition*. John Willey & Sons, 1995.

[96] R.R. Tummalo. *Fundamentals of Microsystem Packaging*. McGraw-Hill, 2001.

[97] F.T. Ulaby. *Fundamentals of Applied Electromagnetics*. Prentice Hall, 2002.

[98] V. Venkatraman and W. Burleson. Robust multi-level current-mode on-chip interconnect signaling. in the presence of process variations. In *Proceedings of the 6th International Symposium on Quality of Electronic Design*, pages 522–527. IEEE, 2005.

[99] B. Victor. Bus encoding to prevent crosstalk delay. Master's thesis, University of California at Berkeley, Berkeley, California, May 2001.

[100] B. Victor and K. Keutzer. Bus encoding to prevent crosstalk delay. In *Proceedings of International Conference on Computer Aided Design*, pages 57–63. IEEE/ACM, 2001.

[101] Xilinx. Virtex fpga family specifications. Technical report, www.xilinx.com, 2005.

[102] B. Young. Return path inductance in measurements of package inductance matrixes. In *IEEE Transactions on Components, Packaging, and Manufacturing Technology*, volume 20, pages 50–55, February 1997.

[103] J.L. Zerbe, P.S. Chau, C.W. Werner, T.P. Thrush, H.J. Liaw, B.W. Garlepp, and K.S. Donnelly. 1.6 Gb/s/pin 4-PAM signaling and circuits for a multidrop bus. In *IEEE Journal of Solid-State Circuits*, 36(5):752–760, May 2001.

Index

Kolibri 360 GmbH
Ferdinand-Jühlke-Straße 7
99095 Erfurt, Deutschland
produktsicherheit@kolibri360.de